W9-BAE-358

Those Wonderful Women in Their Flying Machines

Those Wonderful Women

in Their Flying Machines

THE UNKNOWN HEROINES OF WORLD WAR II

BY SALLY VAN WAGENEN KEIL

RAWSON, WADE PUBLISHERS, INC. NEW YORK

Library of Congress Cataloging in Publication Data
Keil, Sally Van Wagenen.
Those wonderful women in their flying machines.
Includes index.
1. World War, 1939–1945—Woman. 2. Air pilots, Military—United States. 3. United States. Women's Air Service Pilots. 4. Women in aeronautics.
5. World War, 1939–1945—Aerial operations, American.
I. Title.
D810.W7K43 1978 940.54'49'73 78-55606
ISBN 0-89256-066-5

Published simultaneously in Canada by McClelland and Stewart, Ltd.
Manufactured in the United States of America
by American Book–Stratford Press, Saddle Brook, New Jersey

Designed by Helen Barrow
First Edition

FOR THE MEMORY OF MARY PARKER AUDRAIN

"Indeed my aunt's legacy unveiled the sky to me . . ."
Virginia Woolf

Acknowledgments

This is to acknowledge Rockwell Stensrud, whose guidance and vision did so much to shape this book; Edward and Catherine Parker Keil, my parents, and Mary Keil and Jane Keil, my sisters, for their loving support; E. Richard Larson for his friendship and understanding; Valerie Andrews, my sounding board; Harvey Klinger, my agent; the directors of the Columbia University Oral History Collection, for their interest and incorporation of the WASPs' memoirs into the History of Aviation collection; the United States Air Force; the USAF Museum; Colonel W. Bruce Arnold; and finally, ultimately, the many former members of the Women's Airforce Service Pilots across the country who shared so generously of their personal archives and their memories.

Author's Note

While I was growing up in the 1950s, my aunt held a special mystique for my two sisters and me. She was five-foot-eleven, with a ruddy, healthy complexion, loved opera and Beethoven, had a flair with wardrobe and decor and was the director of personnel for a large retailing corporation in New York. But what thrilled us most about her was that she had been a WASP, and had flown B–17s during World War II.

By the early 1972, I had become a writer in New York and she and I had become friends. By then, I was well aware that she was in the vanguard of her generation; I also knew how much she suffered from going beyond the normal pattern for women. Awed by the prospects for us of the new generation, she took as much pride in me and her other nieces (a doctor, a lawyer, a teacher and an MBA) as we had always had in her. "It's very like the war," she once told me, "but you won't have to go home." In the summer of 1973, she died. The only memorabilia she had saved from fifty-five years of life were photographs from her flying years, he graduation certificate from Avenger Field in 1943 and a WASP roster.

I realized then how little she had talked about the WASP experience that had been so important to her. Did she avoid the subject? Had we never asked? Determined to learn what I could no longer ask her, I took her roster and went in search of the WASPs. I found them in one room apartments, hilltop ranches with peacocks on the lawn, oceanside A-frames, mobile homes, suburban split-levels off golf courses and Manhattan penthouses. I listened late into the night to memories long buried. (Sometimes their children sat with me and heard them, too, for the first time.) They talked about what being a WASP had meant to them, and they talked about flying. I laughed with them, I sometimes cried. I also flew with them. By 1976, I had become a pilot myself. The WASPs' spirit, even after thirty years, was irresistible.

In the national crisis of war, women were needed to do the jobs men usually did and were invited into the mainstream of American production. The WASPs—like lots of young women who found themselves taking unexpected responsibilities and opportunities during World War II—learned a sense of legitimacy, accomplishment, usefulness and challenge as professionals, as well as financial freedom and, especially for the WASPs, geographical independence from their families and hometowns. They were also women who took one another seriously as professionals. Often, they literally put their lives in each other's hands. The WASPs today still talk of those years as an epitomized experience in their lives. It is remarkably easy for a woman my age, in these times, to relive their WASP years right along with them.

The WASP story today is one of success, of courage, of humor, but also of scandal. While the WASPs viewed the difficulties imposed on them by the war-era climate as inconveniences to be expected, we identify countless instances of prejudice and discrimination. For them, the hazards and the incomprehension were peripheral to their vision of doing something special, something beyond—to fly. This is what makes the WASPs, to me, such wonderful women. They were, often in spite of themselves, pioneers.

Contents

Those Wonderful Women in Their Flying Machines

I

One of the Best-Kept Secrets of World War II

ONE SNOWY WINTER EVENING, four people, all good friends in their late twenties, sat around a table in a downtown Washington, D.C., restaurant. They were talking about current political events, but at some point, their conversation turned to World War II. Though none of them had yet been born when VJ Day came, they had studied the war in college, read about it in books and magazines, and seen a number of war-related movies and television programs. They had also heard much about those years from their parents.

"Was your father in the war?" asked one of the young men.

"He was in the infantry. He made the landing at Anzio."

"No kidding? You're lucky to be around," said the first. "My father had what they called back then a 'war-essential' job, managing a munitions plant in California."

One of the women at the table began to smile.

"What are you smiling at?" her friends asked.

"My father was an intelligence officer," she said. "But my mother was a pilot. She flew B-25s."

"A woman pilot? During World War II? I didn't know there were any!"

The response was always the same. Yet during World War II, 1,074 American women flew for the United States Army Air Forces (AAF) as members of the Women's Airforce Service Pilots. They were known as WASPs. From 1942 to 1944, young women all across the country responded to a call for women pilots by aviatrix Jacqueline Cochran. Women were needed in the war effort to fly men's missions at home so that men could drop bombs and dogfight abroad.

The Women's Airforce Service Pilots was the most exciting and admired of all women's military units during World War II, and by far the most controversial. Yet after the war the WASPs vanished from national memory. Few of the WASPs themselves have discussed that period of their lives, except, perhaps, with their families. No one else had asked. Nor did the AAF's women pilots receive the national recognition of veterans status for their war service for over thirty years. The price for their extraordinary accomplishment has been silence.

In December 1941, at the time of Pearl Harbor, under two hundred American women had pilots' licenses and the hours of certified flight time deemed necessary by the Army Air Forces for a woman to qualify to be a military pilot. By mid-1944, however, Jacqueline Cochran had received over 25,000 applications from women who had begged and connived their way into the cockpit in order to fulfill the flying requirements to join the WASPs. She selected the 1,830 top candidates, who paid their own transportation to Sweetwater, Texas, to begin six months' training at Avenger Field, the only all-female air base in history.

Some of the women were Park Avenue debutantes, actresses, doctors, models, students, aviators and corporate and political heiresses; famous athletes, like golf pro Helen Dettweiler; the first woman airline pilot, Helen Richey; one of the "Flying Hutchinsons," a celebrated family which had flown around the world in the 1930s; Margaret Kerr, of the Oklahoma political dynasty and Marion Florsheim, of the shoe fortune. But most were simply adventurous young women who saw the boys next door going off to war with great fanfare, and wanted to go, too. The WASP meant travel, danger, self-fulfillment, usefulness, and, above all, the opportunity to take to the air in the world's most sophisticated airplanes. "I wasn't a feminist," said a WASP B–17 pilot thirty years later. "I just wanted to fly."

The WASPs did indeed fly—over 60 million miles—in every airplane in America's air arsenal, from the P–51 Mustang fighter to the B–29 Superfortress. One test-flew the AAF's first experimental jet. The WASPs ferried thousands of airplanes to coastal ports for shipment to war theaters. They served as test pilots for the airplanes flown by AAF cadets still in flight training. They towed targets to train air-to-air and ground-to-air gunners; they flew simulated strafing and smoke-laying missions, radar jamming and searchlight tracking missions. Right along with their male counterparts in the AAF, WASPs performed every vital flying mission in America's massive preparation for war in the air. But they flew longer hours and with a lower accident rate than their fellow men pilots not serving in combat.

The WASPs lived a paradox. They studied aerodynamics and sunbathed at Avenger Field, in blistering, windswept west Texas. They flew hooded takeoffs totally by instruments and received coveted nylon stockings for Christmas at a four-engine bomber school in Columbus, Ohio. They swept alligators off the runways and made curtains for their barracks out of the muslin color-coded bullet-riddled sleeves they towed to train aerial gunners while assigned to Fort Myers, Florida. They urinated into pails through funnels in frigid, unpressurized fuselages and wooed their male navigators in the panoramic turrets of a B–25 Mitchell bomber over Kansas. They founded the "Mile High Club"—those who made love over 5,000 feet. And they sang new words to "Bell Bottom Trousers":

Zoot suits and parachutes
And wings of silver, too
He'll ferry planes
Like his mama used to do.

Even wartime Hollywood, which portrayed the WASPs in a 1944 film called *Ladies Courageous* (starring Loretta Young as Jacqueline Cochran), could not come close to the real women who were flying airplanes for the war effort. The WASPs averaged 5-foot 6 inches, 128 pounds, with 35-inch chests. They passed the military's toughest intelligence and physical exams. At Cochran's insistence, the Air Surgeon General's office, eager to learn how the female system functioned over 10,000 feet and over weeks of tough flying, tested and measured their every move. The results were revolutionary for the era. They proved women could do the job.

In the larger laboratory of society, however, statistics did not count. As meticulous as Cochran was about publicity, the WASPs became the sizzling focal point of a long-smouldering political debate, in the Army and in Congress, about women doing men's jobs.

Commanding General of the Army Air Forces, H. H. "Hap" Arnold, gave Cochran and her women pilots program wholehearted support. The reaction at local air bases was not always so enthusiastic. Together in Washington, Arnold and Cochran scrutinized every report that came across her desk and investigated every rumor. Base commanders did not allow WASPs to be first pilots. WASPs were certified to fly aircraft before receiving sufficient training, to increase their recorded accident rates. These obstacles were routine.

One fall morning in 1943, Cochran learned that a WASP had crashed to her death in North Carolina. When the official report did not arrive, she flew to the base. Headquarters argued, "Heavy paperwork." Cochran's inquiries, however, revealed that traces of sugar had been discovered in what remained of the gas tank. Though stunned, Cochran investigated no further. One WASP had died, but such a scandal would kill her entire program.

In all, thirty-eight WASPs gave their lives flying for their country. But every time a WASP climbed into the cockpit to fly a war plane across the country, she faced possible failure of the new engine or a sudden, engulfing storm. The bullets shot at the WASPs' sleeve targets were real; the mechanics repairing the airplanes they tested, overworked. Often, they were given the flying jobs that men pilots had refused as being too dangerous. But the WASPs had grown up in an era capitivated by spectacular, heroic flying feats and charged with the promise of a world totally changed by the airplane. The romance of flight was not only in the hearts of men. The WASPs, too, braved the risks, to be part of that emerging air-age heroism. In so doing, they were rewarded with the two best years of their lives, and precious memories to be locked into attic trunks.

In 1944, when they were the focus of national debate, columnist Drew Pearson, one of the most vehement critics of the AAF's women pilots, wrote that the WASPs had not yet lost their sting. His words echo across the decades as, a generation and a half later, the WASPs' silence is at last being broken.

II

America's New Frontier: The Air

On the blustery morning of December 14, 1903, two brothers, thirty-two and thirty-six years old, left the coal stove of their cozy wooden shed and stepped out onto the windswept dunes of Kitty Hawk, North Carolina. By mutual agreement, the elder of the two tossed a coin.

Six years earlier, in their bicycle shop in Dayton, Ohio, Wilbur Wright scoffed at his younger brother Orville's suggestion that they try to build a horseless buggy. The newfangled contraption promised to encroach on their bicycle business and they might do well to stay ahead of the times, Orville argued. "Impossible," said Wilbur. "Why, it would be easier to build a flying machine!" So they built a flying machine, and called it, optimistically, the *Flyer*.

For the most propitious location to conduct flying experiments with their machine, the Wrights consulted the United States Weather Service. Assuming that flight would best be served by strong air currents, the Weather Service suggested the desolate Outer Banks of North Carolina. Off shore, the Gulf Stream converged with the larger Atlantic, resulting in raging waters and winds so violent that many a ship had been smashed against the shallow sandbars.

The Wrights took the advice, and that wintry day in 1903, as the winds howled off the surf, the coin hit the sand. Wilbur Wright had won the chance to be the first air pilot—ever.

The *Flyer* was in readiness. Its 200 pound, 12 horsepower motor, adapted by the Wrights from a Pope-Toledo automobile, was in perfect working order. The bicycle chains, criss-crossing behind the pilot's pallet on which Wilbur would lie, were securely fastened to the two wooden propellers. The rubber tubes, which would carry water to the radiator, were tight. The second-growth ash wooden spars and uprights were solid. And the glued cloth was flawlessly sewn and stretched around the wing structure to form two forty-foot membranous strata, six feet apart, which would lift Wilbur over the dunes. Trusty bicycle cement held the rivets fast.

The *Flyer*'s sledlike runners rested on the starting end of two parallel wooden rails 100 feet long (these bicycle builders knew well not to trust rubber tires on takeoff). Wilbur nestled his chest onto the pilot's platform and gripped the wing-warping and rudder levers. Orville started the motor, the propellers began to spin and the *Flyer* darted forward along the wooden rails.

In spite of their long scientific preparations, however, history was not about to be made that day. Being America's first air pilot, Wilbur Wright's flight instruction had been somewhat haphazard, limited to a few exhilarating but precarious flights in the Wright's experimental gliders. Like many beginning pilots who followed him over the decades, Wilbur pulled up the front of his flying machine too high as it lifted off the rails. The *Flyer*, like an underpowered automobile on a steep hill, slid back with a thud to the ground.

On December 17, after three days of regluing and reriveting, it was Orville's turn. The brothers flew a white flag over their shed to signal the other residents of their thin, sandy strip of land—members of the Kill Devil Hills Life Saving Station—to come and be witnesses. When their neighbors arrived, one was enlisted to push the shutter button of a large black camera pointed at the end of the takeoff rail.

Shivering with cold and excitement, Orville donned his automobile touring cap and goggles and lay down on the pilot's pad. As the motor hummed and the propellers twirled, the *Flyer* lurched and skidded along the rails as Wilbur ran alongside to steady the wing. Suddenly, the *Flyer* shot upward and Wilbur was left with his arms in the air. The photographer snapped. The *Flyer* flew. Orville scrambled with the levers as the flying machine soared and dipped through the air like a dolphin. On one of its aerial plunges, it met the sandy dune and stopped. Man's first sustained flight in a heavier-than-air, motor-powered flying machine had lasted twelve seconds.

The news of man's conquest of the air was broken by the telegraph relay operator in Norfolk, Virginia, to the local paper, aptly called the *Virginian Pilot*. The Wright's flight made the front page. But for almost a month, the rest of the world remained ignorant of the new age it had just entered. In mid-January, the New York *Herald Tribune* collected information from various sources—including a letter of explanation from the Wrights which was all but ignored—and published an in-depth, illustrated article on man's first powered flight. The artist's conception of the *Flyer*, however, was closer to a Jules Verne creation, and the writers seemed more absorbed in the ancient Daedelus and Icarus myth than in the Wrights' actual accomplishment.

In May 1908, Wilbur Wright sailed for France. At last, the Wrights had received a prospective licensing offer from a French financier to manufacture *Flyers*, on the condition that Wilbur could make two flights of fifty kilometers in the same week. To ensure the venture, the brothers engaged a sales representative, Hart O. Berg, who had introduced the American submarine and electric automobile to Europe. For demonstrations, Berg found an ideal location, convenient to mechanical shops and to a forward-thinking audience—the Hunaudières race track in Le Mans.

To Berg's dismay, however, the French crowds were skeptical. Calling forth all of his promotional expertise, he conceived of a stunt which would surely arouse attention and inspire confidence in the dependability of the Wright *Flyer*. He sent his wife down from Paris to Le Mans to take a flight.

Edith Berg was delighted. She tied her long skirts around her knees and sat, legs stretched out in front of her, amid the scaffolding of canvas and piano wire. (The Wright *Flyer* now carried two, decorously sitting up.) Wilbur Wright, as always in his high starched collar, grey suit and touring cap, climbed into the seat beside her.

The machine was fastened by a wire to a large rock. After two men got the 30-horsepower motor going and the two propellers spinning, they unhooked the wire and the *Flyer* shot five stories straight into the air. Edith Berg squealed and in her enthusiasm almost let go of her seat. A two-minute, three-second flight over Le Mans was long enough to prove their point, and all too soon, Edith Berg's feet reluctantly returned to the mud of the race track. But she had made history. She was the first woman to go up in a flying machine.

Telegrams of congratulations poured into the Bergs' Paris residence which celebrated Edith Berg's courage and the Wrights' marvelous machine. The *journaux* were ecstatic. But in the Paris *salons*, they whispered, "That crazy American woman! And imagine her husband's letting her do it!" Flying, they said, was even more shocking than driving a motor car.

Over the next few years, the Wrights in Dayton and other air pioneers like Glenn Curtiss in upstate New York, trained a growing number of adventurous aeronauts—as pilots were then called—who continued to experiment with flight. Among them were Houdini and a young Army officer named Henry H. Arnold, and to everyone's surprise—women.

On August 1, 1911, Harriet Quimby, a twenty-seven-year-old writer for *Leslie's Magazine*, a *Time* or *Newsweek* of the era, appeared with her gossamer bi-plane at the Long Island headquarters of the Aero Club of America, which licensed pilots before the government accepted the responsibility in 1925. The Aero Club was confounded. A woman had never requested to be licensed as an aeronaut before. Quimby fixed her dazzling green eyes on theirs and politely suggested that club members should at least let her demonstrate her flying ability.

As club officials looked on dubiously, Harriet Quimby and her aeroplane shuddered off the ground and then soared over a nearby potato field. She banked gently, and flew back toward her starting point. As officials watched in amazement, Quimby set her machine down within eight feet of where she had begun her flight. She had just broken a club record for landing accuracy. "You have to give me a license, don't you?" she said. And they did, though as amazing as her flying expertise was the plum-colored satin flying suit she wore, with hood, knickers and puttees—all of the same material. "Flying seems easier than voting," Quimby quipped to admiring reporters. The Nineteenth Amendment was not ratified until 1920.

On July 2, 1912, as Woodrow Wilson was being nominated for President of the United States, an aero meet crowd of more than 1,000 peered over Boston's Dorchester Bay awaiting the return of Harriet Quimby from a twenty-mile speed trial to Boston Lighthouse and back. Everyone sighed as the raven-haired woman aeronaut in her pure white monoplane—single wing flying machines were rare in those days—swept gracefully over the crowd and headed out into the red-orange sunset now illuminating Dorchester Bay.

Suddenly something went wrong. The white tail flashed upward and then dropped perpendicular to the bay. The machine stood on end, hung suspended for a moment, then began to plunge, tail down, toward the water. Quimby's tiny body seemed to pop out of the plummeting plane. The crowd watched in horror as she tumbled over and over, outlined against the flaming sky, and splashed into the water. The white monoplane floated down after her, hit the waves nearby and, rocking forward, buried its nose in the depths.

The next day, Quimby's fellow aeronauts tied black crepe streamers to their stay wires and flew on with the aero meet. Quimby's accident was never explained. But in the wreckage of Quimby's flying machine, aero meet officials found an aneroid barometer. At one point during her flight, it had registered an altitude of over 5,000 feet. Harriet Quimby had flown higher than any woman had ever flown, before she plunged to her death.

SIX YEARS after Harriet Quimby's death and Woodrow Wilson's election, America emerged triumphant from the Great War, having matched its powerful motherlands of Europe in valor and technical competence. Those who survived the mortar fire, mustard gas, cold and disease in the trenches may have won the Great War, but far more spectacular heroes were made of slightly over a hundred men: those who fought high above the mud and boredom, those who battled the aristocracy of Europe in the air, the boys from democracy's cornfields and cow pastures—the flying fields of America. For America on a postwar high of power and pride, a sensational, novel and peculiarly modern hero had emerged.

The Ace. The term was first used by the French Armée de L'Air, meaning the top of the pack. Touted and bemedaled by every nation, he was partly used as a morale booster. But, though helping the world forget the inglorious mess of the trenches, the exploits and trappings of the Ace *were* the least warlike imaginable. He wore jodhpurs and dashing helmet and boots, a white scarf to warm his neck in the brisk winds aloft and wipe his goggles.

Glamorous as he might have been, the Ace was flung into aerial warfare in the most primitive of conditions. The maneuverable biplanes and tri-planes turned on a dime, their inmates' stomachs either sinking through their seats or forced up in their mouths while they dove and circled, defying gravity, as if on a three-dimensional roller coaster, a mile over the heads of the trench-bound doughboys. In the midst of the roaring furor, there might be an explosion of flame as a bullet hit a fuel tank. Then a flier would fling himself out of the cockpit and free fall to earth; he preferred a quicker, cleaner death than being roasted alive. The World War I Ace had no parachute. The seatpack, all thirty-five pounds of it, which was the constant companion of the World War II pilot, was not put into use until 1921.

The earliest World War I warplanes were called scouts, or observer planes, and were employed to get a better view of enemy positions. None was armed, but reconnaissance pilots soon could not resist grabbing a twenty-pound bomb and throwing it out of the open cockpit onto an enemy installation like an oversized dart at a pub dartboard. At British Royal Flying Corps bases, the first "fighter" pilots of 1914 ran from their canvas-tent barracks to their canvas-winged aeroplanes on missions—to ram any Zeppelin approaching British shores.

The newly invented machine gun, attached in front of the pilot, was soon to create more deadly air combat. In 1915, twenty-five-year-old Dutch-born Anthony Fokker invented for the Germans an interrupter gear that kept the bullets from hitting the rotating propeller blades.

In fact, in four short years, machines and strategy had developed which would lead directly to a major air war years later. The scout, in which the early pilot suddenly found himself nose to nose with an enemy observer at 500 feet and in consternation shot his pistol, became the prototype of later pursuit planes. By 1918, monstrous metal and plywood bombers had been built with four engines and 100-foot canvas wings, as broad as the future Flying Fortress, which could carry 1,650-pound bombs and hold several gunnery positions. The Germans (young Hermann Goering among them) had developed a bombing strategy, covering all of southeast England, which would presage the devastating raids of the Battle of Britain just two decades later.

At the end of the war to end all wars, however, a thousand aeroplanes were burned by the United States Army and the rest were sold for pennies. Few of America's top Aces lived to see the end of the war (nor did the Aces of any other combatant nation). But the World War had produced hundreds of freshly trained pilots who would help usher in the golden age of aviation and keep the Ace image alive. After the 1918 armistice, the thrill of winning a world war became synonymous with the thrill of man in flight.

As THE BARNSTORMERS of the 1920s (many of whom were World War I-trained) wielded their magic on spectators across the country, they introduced the wonders of flight to thousands of Americans. Crowds paid to gawk at him walking out on the wing of his plane to retrieve his handkerchief, dangling from his landing gear, or exchanging cockpits with another flier at 1,000 feet. He rigged his bi-plane with smoke devices and staged mock dogfights which were almost as dangerous as the real thing over Verdun or Ipres. For his crowds, that he flew was miraculous; that he crashed to the earth—curse you, ripped wing fabric! curse you, twanged rudder wire!—was to be expected.

Unlike his crowds, the barnstormer knew that science held him up. He was Icarus insouciant of any fall. Not made of wax and feathers, his wings worked—they were shaped for flight. The top of each wing was curved, the bottom was flat. Takeoff, that moment when his plane was lifted from the ground, was a simple phenomenon. He was merely going fast enough so that the air flowing over the curved top of his wings, having farther to go, was streaming by faster than the air flowing past their flat bottoms. The faster air immediately above the wings was less dense than the slower air below, and the wing was literally pulled into the vacuum and up off the ground, bringing him and his machine with it. He could soar and loop circles perpendicular to the earth, climb high among the clouds and fall spinning like a pinwheel toward the ground. His gasping crowds thought he was playing with his life, when he was merely playing with the air and the various control surfaces of his aeroplane.

Flying was not magic, it had become scientific fact, as well as a new dimension. But few people flew. Fewer still built aeroplanes. Though names to be prominent two decades later were in the picture—William Boeing, Glenn Curtiss, Glenn Martin, Lawrence D. Bell, James McDonnell and Donald Douglas—others now forgotten, like the Granville Brothers and Charles Kinner, were equally active in what was still a cottage industry.

Though two decades had passed since the Wright brothers' patent was approved in 1906, few yet dared invest dollars and sweat in such a fantasy as flight. But those who did, and survived, were idolized.

In 1927, Gene Tunney was champ, Isadora Duncan went to her glory, the jazz age was in full swing, and the in-crowd filled the speakeasies of Prohibition America. Then in May, "something bright and alien flashed across the sky. A young Minnesotan who seemed to have had nothing to do with his generation did a heroic thing . . ." So fellow Minnesotan F. Scott Fitzgerald described an event which became the glorious symbol of a new age: a twenty-five-year-old air-mail pilot named Charles Lindbergh, Jr. took off from New York, flew thirty-three hours alone across the Atlantic Ocean, and landed in Paris.

"L'Americain est arrivé!" shouted newsvendors along the Champs Elysées, the night of May 22, 1927. The next day the news arrived in America. On May 23, in Indianapolis, Indiana, seven-year-old Betty Phillips, too, heard shouts. Rushing to her front door, she saw a newsboy selling "extras." Up and down the street her neighbors were out on their porches. He offered her a dime to take her side of the street. When she took her stack of papers, she had goosebumps.

After his flight in the *Spirit of St. Louis* on the moist breath of the wind, Lindbergh was borne home on the broad shoulders of a huge U.S. Navy cruiser sent by President Coolidge, who declared June 11 "Lindbergh Day." (The trip back took four times longer than his sensational flight.) For over three weeks, the first four pages of the New York *Times* covered nothing but Lindbergh. Detroit named an avenue, Texas named a town, the Pennsylvania Railroad named a train, San Diego named an airport, and the Brooklyn Zoo named a baby elk—all after Lindy. And in the era when the business of America was business, Wright Company stock soared and Wall Street gave Charles Lindbergh the grandest tickertape parade it had ever fluttered.

When Charles Lindbergh flew the *Spirit of St. Louis* over Niagara Falls on his way to Dayton to salute Orville Wright (Wilbur had died of typhoid in 1912), among the cheering crowd was another little girl, six-year-old Marion Hanrahan, who gripped the hand of her aunt as tightly as she could.

In 1935, when Marion Hanrahan was fourteen, she would spirit sheets out of her mother's linen closet, cut her New Jersey high school classes and steal out to Bendix Field (now Teterboro Airport). Patching airplane wings would earn her flying lessons. The flying community was still so small that the world's most famous woman aviator, Amelia Earhart, a frequent visitor to Bendix Field, would sit leisurely giving the teenager flying tips.

But during the Lindbergh era, fewer than one percent of all Americans had been anywhere near an airplane. Some of Marion Hanrahan's youthful contemporaries were among the lucky, and daring, one percent.

When Dolores Meurer was three years old, she and her mother passed a St. Louis street photographer who was taking pictures of children sitting in a tiny wooden airplane. Her mother combed her blonde Dutch-boy and lifted her into the plane; the photographer snapped. The picture remained Dolores' favorite during her childhood. When she was six, she was taken out to Lambert Field, the

St. Louis airport, where her father spent the extravagant sum of $2.00 for the two of them to take a ride in a barnstormer's bi-plane. Twelve years later, Dolores Meurer was washing down and fueling airplanes at the St. Louis airport in return for flying lessons.

Even those growing up during the 1920s and 1930s whom the barnstormers did not touch, however, could not avoid the excitement of a world being changed by aviation. As a child in Jamestown, North Dakota, Vivian Gilchrist would grab the newest adventure books from the boys' section of the public library, featuring the daring exploits of Don Sturdy and Tom Swift, and climb into her favorite tree to read. Soon, Vivian's adventure books were supplemented by her parents' newspaper, which fired her with adventures even more marvelous, and done by real people. In 1924, two Army lieutenants flew tiny bi-planes all the way around the world. In 1925, Admiral Richard E. Byrd flew over the North Pole, and in 1927 explored the South Pole in an airplane. There was, of course, Charles Lindbergh's solo journey over the Atlantic. But the next year, a spectacular thing happened. The Atlantic was actually flown by a *girl* named Amelia Earhart. Though she had just been a passenger, it seemed to Vivian quite an adventure, as it did to the rest of the world. To follow were Englishwoman Amy Johnson's solo flight from England to Australia in 1930, Wiley Post's around-the-world flight in 1931, Earhart's solo Atlantic crossing in 1932.

From California to Cape Cod, little girls, along with their brothers, caught the spirit of freedom and limitless possibilities now that human beings could fly. In the foothills outside Sacramento, Isabel Steiner made model airplanes until so many were suspended from the ceiling a grownup could barely enter her bedroom. Three hundred miles down the San Joachin Valley, Charlotte Mitchell played pilots and parachutes with a little girlfriend. In Birmingham, Alabama, all seven-year-old Nancy Batson wanted for Christmas in 1929 was a flight suit, helmet and goggles.

Lindbergh and his heroic compatriots of the air gave the American frontier a new dimension. Even hometown streets became magical from on high. The earthbound had gained a new perspective and mobility. Young Americans were worshipping a new breed of heroes. These modern heroes had all the qualities of legendary ones who braved the unknown, alone. But the heroes of the air symbolized the new century. They had a unique combination of gallantry-of-old and mastery of the new technology. Modern technology not only created their heroics, but also proliferated their fame on film; these heroes had higher public visibility than any before them. And as airmail and passenger routes began to connect America's towns and cities, the daily hum of an engine seemed to sing, "You can be up here, too."

This twentieth-century breed of heroes redefined heroism. Those impossible feats of long-distance solo flight, those awesome, birdlike aerobatics, demanded a courage not of brute strength and struggle of mortal combat, but of refined intellectual mastery, endurance and spirit. The modern hero of the air could be, and often was, a woman.

III

Amelia Earhart: A Different Kind of Heroine

In April 1928, less than one year after Charles Lindbergh flew the Atlantic, a twenty-nine-year-old social worker in a Boston settlement house was reluctantly called away from her group of children. A man on the telephone said his business with her could not wait.

"Miss Earhart, would you like to participate in an important flight for the cause of aviation?" asked Captain Hilton Railey in a smooth Southern-accented voice. "It's likely to be hazardous."

That afternoon, accompanied by the directress of the settlement house as chaperone, Amelia Earhart was shown into Captain Railey's public relations office in downtown Boston. Though the office was comfortable and handsomely furnished, she was suspicious of his offer. She had been demonstrating Kinner Canaries, small yellow biplanes manufactured by a friend of hers from California, where she had learned to fly in 1920, and was active enough in aviation to be elected vice president of the Boston chapter of the National Aeronautic Association. Since she was known in the area, such calls as Railey's usually were to ask her to fly illegal liquor into the country from Canada.

Railey, she soon learned, was in a different league. "How would you like to fly across the Atlantic?" he asked. The flight was to be sponsored by a wealthy Pittsburgh heiress named Amy Phipps Guest, now of New York and London, whose husband, Frederick, had been Air Minister under Lloyd George. Mrs. Guest had bought a trimotored Fokker and named it the *Friendship*. Her ultimate purpose was to have a woman fly the Atlantic.

The *Friendship* was already being fitted with pontoons; its fuselage was being painted red and the wings gold for high visibility if downed at sea. Two World War I veterans, Ace pilot Wilmer L. "Bill"

Stultz and mechanic Lou "Slim" Gordon, were preparing a flight plan and taking meteorological readings for a proposed route. All that was needed was to find "the right sort of girl" to make the flight. If she were a flier herself, all the better.

As introduction, Earhart gave Railey a verbal resumé of her flight experience—500 hours of solo flying in the Northeast and California, and a women's altitude record of 14,000 feet, which she had set in 1922. Railey's questions were polite but seemed perfunctory: "What is your education—if any?" he asked. "How strong are you?" Railey, in truth, could barely contain his excitement. Though he had screened other candidates, he knew from the first moment he saw this one that she was perfect to be the first woman to make a transatlantic flight. Amelia Earhart, with her fair dune-grass tufted hair, tall, lanky grace and frank smile, looked enough like Charles Lindbergh to be his sister. Here was the "Lady Lindy" Railey was looking for.

The choice of Amelia Earhart to make aviation history was settled when she sat before a panel of three men in New York the following week: Mrs. Guest's attorney; her brother, representing the Phipps family; and the man who was the initiator of the entire venture, a New York publisher and friend of Railey's from World War I, George Palmer Putnam, whose most recent triumph had been *We*, by Charles Lindbergh. "I would hope," Earhart told the panel, "that I might have some time at the controls."

Two days later, she received a contract in the mail. The pilot would receive $20,000 for the flight, the mechanic, $5,000. Amelia Earhart was along for the ride—a ride that would change her life.

The North Altantic, which had been flown successfully only twice, threatened among the biggest, most durable storms in the world. Flying 2,300 miles through one was not something a pilot or passenger should look forward to, especially when the metal joining the wings to the fuselage of 1928-model airplanes had not been tested for such stress and radio equipment or flight instruments could fail at any minute. Once shore was abandoned, radio signals to confirm location came only from whatever ships might be lumbering below the flight route.

Captain Railey's wife, in a woman-to-woman chat, assured Earhart she need not be irrevocably committed to such a perilous undertaking. "When a great adventure is offered you," replied Earhart, "you don't refuse it, that's all." Nevertheless, she wrote letters to each of her parents (she called them her "Popping-off Letters") to be opened in the event she did not survive the voyage. "My life has really been very happy," she wrote her mother, "and I didn't mind contemplating its end in the midst of it."

On June 4, 1928, the flaming red, tri-motored *Friendship* took off on its golden wings from Boston Harbor toward Newfoundland, the jumping-off point for the 2,300 mile nonstop transoceanic flight to Britain. Amelia Earhart had taken three weeks' vacation from her job at the Boston settlement house. Vowed to secrecy by the organizers of her flight, she had kept her word so well that she had not even told her mother and sister, now living with her in Boston. They read of her departure in newspaper headlines: "Girl Flyer Braves Atlantic."

Much to the dismay of organizers and crew alike, storms raging either ashore or in the mid-Atlantic kept the *Friendship* grounded in Newfoundland for two weeks. Finally, late in the afternoon of June 17, the weather cleared long enough for optimism. Loaded with scrambled-egg sandwiches, a thermos of coffee for the pilot, five gallons of water, oranges, chocolate, a bottle of malted milk tablets, one dufflebag shared by the three occupants and 700 gallons of gasoline—just enough to make Southampton, England, where their patron, Mrs. Guest, awaited them—the three-ton *Friendship* took off eastward. It needed four takeoff passes down the waters of the harbor before finally lifting off and allowing the seventy-two foot wingspan to bear its load.

Once airborne, Earhart knelt between two extra gas tanks in the darkened, almost bare cabin, as she took pictures out the tiny window and wrote the flight log, her job during the next many hours. She also spent time squatting behind the experienced former Ace, watching how he maneuvered the *Friendship* by instruments through the disorienting fog. Earhart was apprenticing his every move.

Hours passed. In the early morning, Earhart's log, which had tended to poetry on cloud formations, began to change its tone. The fog had become a storm. Water was dripping in the cabin window. Without visual reference to the stars and only forecasts of winds aloft to trust, there was no assurance they were on course, which, because weight had strictly limited their gas supply, had to be almost exactly as planned. The radio went dead. As they started their descent, one of the three motors began to cough, then the other two. After using more gas than projected to fly around the worst of the storm, they had less than an hour of fuel left.

Suddently, at 3,000 feet, they broke through the clouds. On the horizon, bright with morning sun, was the jagged outline of land. Twenty hours and forty minutes after leaving Newfoundland 2,246 miles before, the *Friendship* splashed down in a small harbor outside Burry Port, Wales. With no radio, storms all the way across the Atlantic and faltering engines, pilot Bill Stultz had landed within one mile of his planned line of flight to Southampton. Only fuel restrictions had robbed him of 150 miles and complete success. Earhart had apprenticed a pro.

The tiny Welsh fishing village of Burry Port hardly expected an airplane to drop from the sky into its harbor. Arrangements for the *Friendship* to be towed in to shore took so long that Captain Railey had time to arrive from Southampton by flying boat. As he rowed alongside the tri-motor, bobbing in the harbor on its pontoons, he met Earhart sitting Indian-style in the doorway of the fuselage. "How does it feel to be the first woman to fly the Atlantic?" he shouted exuberantly in greeting. "Aren't you excited?"

"It was a grand experience," she said calmly. But she had not flown the Atlantic, Bill Stultz had, and he had done a splendid job of it. "I was just baggage," she added. "Someday I'll try it alone."

Regardless of how she felt about her role, Amelia Earhart became an instant celebrity. For the next two weeks, she was feted as "Lady Lindy" by London society. She chatted with Churchill, danced with the Prince of Wales and was asked a hundred times for "a great big smile, please" by photographers.

A high point was meeting Lady Mary Heath, who had puddle-jumped her small, open-cockpit Avro Avian from Cape Town, South Africa, to London—12,000 miles by herself. Earhart was delighted with the chance to express her admiration and was even more so when Lady Heath offered the Avian for a spin. Earhart was so excited with the plane, if not with being at last behind the controls again, that she bought it.

Earhart could not shake the feeling of being a "false heroine." Upon her return to America, Captain Railey arranged a three-city tour, during which, much to Earhart's chagrin, she eclipsed her *Friendship* pilot and mechanic as they faced the same crowds that had adored Lindbergh the year before. But back in New York, Earhart took delivery of Lady Heath's 881-pound Avian. Amid the African town emblems which covered the side of the cockpit was a message: "To Amelia Earhart, from Mary Heath. Always think with your stick forward."

Also awaiting Earhart in New York was a new life. George Palmer Putnam took over from Railey to become agent, manager and publisher of her first book, *20 Hours 40 Minutes,* which she wrote that fall of 1928 from her log notes. In a matter of months, her unpaid transatlantic flight had compensated her well in fame. Ray Long, canny editor of *Cosmopolitan* magazine, who was then publishing Ring Lardner, Ernest Hemingway, Michael Arlen and other writers of the 1920s, invited her to contribute a monthly column as "aviation editor." From advertising endorsement offers, she gleaned over $50,000, with which she began to prepare herself for a future flight, one she had to make, across the Atlantic alone.

Amelia Earhart's apprenticeship with Bill Stultz was only the beginning. To gain navigational experience, she flew her new Avian by herself back and forth across the country, which at the time had few airfields. Once, lost and out of daylight, she landed down the middle of deserted Main Street in Hobbs, New Mexico. Technical proficiency was imperative as well. With Bernt Balchen, a colleague of Commander Byrd's during his Polar flights, Earhart continued her study of "blind flying" by instruments and, for knowledge of the meteorology of the North Atlantic, she haunted the U.S. Weather Service. "Someday I will redeem my self-respect," she had told Railey after the *Friendship* flight. "I can't live without it."

By the fall of 1928, twenty-year-old Betty Huyler had decided she wanted a professional career and entered the nurses' training program at Columbia Presbyterian Hospital in New York City. She had also fallen in love with a dashing naval aviator who was proving to be elusive. Her competition was not another woman, however. It was an airplane.

One November day, a fellow nursing student gave Betty the new issue of *Cosmopolitan* magazine to read while off duty. In it was an article by the new aviation editor, Amelia Earhart, entitled "Try Flying Yourself." When Betty Huyler closed the magazine, she had a gleam of resolve in her arresting green eyes. "Why not?" she thought.

For the next few months, Betty broke speed records in her little car over the Long Island motor parkway from the hospital to Roosevelt Field, Lindbergh's departure point for Paris the year before, to get in a half-hour flying lesson. Then she would race back for hospital duty. The gleam in Betty's eye had become a flame of enthusiasm.

Just after she had earned her pilot's license, Betty Huyler was offered a flying job at Curtiss-Wright Corporation's aviation center on

Long Island to demonstrate for flight school customers flying curricula and a new trainer airplane called the American Moth. Nursing was no longer the only possible career for her.

Betty Huyler stood resolutely by the superintendent of nurses' desk. "I don't think you are going to like that," the older woman told her with a frown. "Any time you want to come back and have a career, you can be a nurse." But Betty wanted to be a pilot.

In November 1929, Betty Huyler and several women pilots from the New York area decided it was time for a national organization. On November 2nd, over twenty licensed women pilots arrived for a meeting at Curtiss Airport. Among them was Betty's inspiration, Amelia Earhart. High on the order of business was choosing the name of the organization. All the names they could think of—"Lady Birds," "Homing Pigeons," "Angels' Club"—sounded straight out of Will Rogers' press reports during the first all-woman air derby held that year. Finally, Earhart's quiet voice interrupted the discussion. Why not name it after the number of charter members? she suggested. A letter went out announcing the formation of a women pilots group. Ninety-nine women pilots across the country responded, and the organization was named the Ninety-Nines. Amelia Earhart was elected its first president.

Twelve years later, Betty Huyler Gillies was married to her naval aviator, had three children, and was herself president of the Ninety-Nines. A utility pilot with over 1,000 hours of flying experience at Grumman Aircraft, where her husband was an executive, Betty was now advocating to her membership, which numbered over 500, that they upgrade their pilots' ratings and get more experience in case they were needed to fly when war came. She would soon be one of the first to be called.

THE EARHART-LINDBERGH duo was the culmination of the modern androgyny of air heroes; in their unassuming, eloquent charm and open, fair-haired good looks, combined with a courageous and trustworthy competence, both were embraced and held close to the public heart. But the rift between men and women participating in the aviation industry was already forming. In 1929, Earhart was asked by Transcontinental Air Transport, the forerunner of Trans World Airlines, to join Lindbergh as a consultant. Lindbergh was chairman of their technical committee; Earhart was given a marketing job, as advisor on women's sales resistance. Lindbergh flew, researching air routes across the country. Earhart toured the country as a passenger, interviewing other women in the plane, and lecturing to women's groups about how safe and pleasant flying was. The fledgling commercial aviation industry, of which Earhart was reigning queen, considered her to be a symbol, as the first stewardesses would be in 1930, that flying was so safe that even a woman would do it.

When Transcontinental Air Transport began forty-eight-hour coast-to-coast air-rail passenger service in July 1929 (passengers flew during the day and took a Pullman all night), the company dramatized the event by having Lindbergh himself fly one of the first routes. Behind him in the passenger compartment of his Ford Trimotor were Amelia Earhart and Anne Morrow Lindbergh, also a pilot. Transcontinental Air Transport had made its point.

By 1932, with the help of George Palmer Putnam, now her husband as well as her manager and publisher, Amelia Earhart's name was, as biographer John Burke states in *Winged Legend,* a "household word." Like many of today's sports heroes, her picture was on billboards across America; her name was on merchandise ranging from luggage (light, for air travel) and pajamas, to a line of buttons and belt buckles made from tiny airplane motor parts.

She had also bid farewell to her 80-horsepower Avian and bought a 500-horsepower Lockheed Vega, the type of plane in which Wiley Post had encircled the globe in eight days, fifteen hours, fifty-one minutes, in June 1931. Like the *Friendship,* the Vega was painted bright red for ultra-visibility if downed at sea.

On May 20, 1932, five years to the day of Charles Lindbergh's departure from nearby Long Island, thirty-four-year-old Amelia Earhart, in jodhpurs, plaid sports shirt and well-worn leather flying jacket, took off alone from Bendix air field in New Jersey for Newfoundland. After four years of preparation, she was ready to make the flight that would legitimize her role as First Lady of the Air, not for others—her adoring public expected nothing more than what she had done—but for herself. During the nexty twenty-four hours, she would have the chance to earn every inch of her lofty reputation.

Night had fallen when she took off from Newfoundland across the broad back of the North Atlantic. She climbed through the clear night sky to 12,000 feet and was speeding along, she calculated, at 180 miles per hour. The *Friendship* on its flight averaged a doddering 113. Suddenly, however, her altimeter needle began to rotate wildly; a key instrument, she realized with alarm, was gone. As long as she had the visual reference of the waves below she would be all right. But the North Atlantic skies did not cooperate. Around midnight, she was swallowed by a storm. Downdrafts and updrafts were so severe that she no longer had any idea what her altitude was. After fighting, white-knuckled, with the controls through an entire hour of rain and lightning, she spotted the moon through a brief cleft in the clouds and decided to climb above the storm. She pulled back on the stick and climbed for several minutes, but felt the controls become sluggish, as if the ailerons and elevator were flapping in pudding. Without an altimeter, she had not realized she was so high that ice was forming on the control surfaces until she saw slush appear on her windshield. Her tachometer needle, which registered the revolutions per minute of the engine, began to whirl as her altimeter had done, and suddenly the plane, heavy with ice, went into a spin. Down she spiraled through the clouds, frantically moving the ineffectual ice-caked elevator, rudder and ailerons. After interminable seconds, during any one of which she expected impact with the water, the warmer air melted the ice and she felt the controls take hold and the plane zoom level again. Below her, less than one hundred feet away, she saw the white caps of the Atlantic. Preferring to fly blind in a calm fog layer than flirt with the turbulence above the waves without an altimeter, she climbed back into the muck. She took a gulp through a straw from a can of tomato juice.

Her troubles were not over. Out the side of her engine cowling, she spotted flames shooting from the exhaust manifold pipe. Then the pipe began to vibrate, and smarting gas fumes drifted into the cockpit. Her eyes began to tear so badly that she could barely see the instruments, those that were left. Flames in the exhaust manifold meant that not all her gasoline was being burned in the engine, but was escaping; her engine could blow up any minute. She abandoned all hope of making Paris, to duplicate the flight of her illustrious male predecessor, and banked onto a compass course straight to the nearest land—Ireland.

At 1:45 P.M., fifteen hours nineteen minutes after she had left Newfoundland on the other side of the Atlantic, Amelia Earhart settled her Vega down into a meadow on Patrick Gallagher's farm outside of Londonderry. A farmhand came running across the field and stared at her open-mouthed as she climbed exhausted and dazed from the cockpit. After several seconds in which they looked at one another like visitors from different planets, she smiled weakly and said, "I've come from America."

George Palmer Putnam boarded the next ocean liner and again,

Amelia Earhart was received not only in England, where she stayed as the guest of American Ambassador Andrew W. Mellon, but all over Europe, with the wildest of accolades. Although bewilderingly excessive, Earhart felt that these tributes, at least fundamentally, were earned. She floated through them, glad enough to be alive. (A postflight inspection of her Vega revealed that the brackets holding the wings to the fuselage were cracked to the yield point.) But at last, the heroic qualities assumed by others to be hers—the courage to be adrift for hours in a voidlike sea of fog without reference to any of the earth's outlines, the almost religious confidence in oneself that in the split second of emergency, judgment will be accurate and save one's life—she had now proven to herself.

ACROSS AMERICA, Earhart's courage and professional success inspired many scientifically inclined young women to struggle for knowledge and a place in aviation. When a fifteen-year-old Californian, Iris Cummings, finished swimming the 200-meter breaststroke and three-stroke medley at the 1936 Olympics in Berlin, her mother asked her, "Now that your competition is over, is there anything you would like to do?" There certainly was. As a reward for two years of training at the Los Angeles Athletic Club, Iris was taken to watch an air show of the most advanced soaring and powered aircraft flight maneuvers in all of Europe.

Iris had not forgotten attending the National Air Races, America's most prestigious annual aviation event, which was like barnstormers' heaven. When she was twelve, her parents took her to Mines Field, now Los Angeles International Airport, then but a square mile of grass with five hangars and a small control tower, to watch the bi-planes loop and twirl above her. Amelia Earhart had raced that year, the first time women had entered the competition.

By the time she was seventeen, Iris Cummings was enrolled in every aviation course offered by the University of Southern California and was the one female member of the university aviation fraternity. Most thrilling, she found herself performing her own air show over the grassy expanse of Mines Field, as a student pilot in a government-sponsored training program. With the advent of World War II, Iris would be one of the first women to fly U.S. Army fighter aircraft, the most advanced airplanes in the world.

TO AMELIA EARHART, flying was more than a test of her own personal worth and courage. It was her love. As First Lady of the Air, Earhart's role as the muse of aviation redoubled as she increased her lecturing activities over the next many months (she delivered 136 in 1936 alone). In fact, her inspiration reached the highest of American homes, the White House.

In April 1933, Earhart and her husband, George Palmer Putnam, were dinner guests of the Roosevelts, newly arrived in Washington. After dinner, Earhart and Eleanor Roosevelt chatted about flying. Earhart was extolling the beauty of flying at night, how the world becomes serene and glittering far above the lighted outlines of the city. Mrs. Roosevelt seemed so enthralled that Earhart, with characteristic forthright spontaneity, asked if she would like to take a flight. Now? asked Eleanor Roosevelt. Why not? asked her dinner companion.

Without telling their husbands, they grabbed a delighted woman reporter, slipped out of the White House, and commandeered a limousine which sped them to National Airport where Earhart arranged to borrow an Eastern Airlines plane. Still wearing her white kid gloves, the First Lady of the Air took off and flew the First Lady of the nation up over the Potomac and Tidal Basin and into the clear night sky over Washington, D.C. for Mrs. Roosevelt's first view of the glittering Capitol from the air.

When they landed back at the airport, Eleanor Roosevelt was so delighted by the sensation that she asked Earhart if she would teach her how to fly. Much to the disappointment of the two women, the President vetoed the whole idea. But Mrs. Roosevelt's interest and faith in women in aviation did not wane. She became their champion when, with the advent of war in Europe, women were officially overlooked by her husband's military advisors as they mobilized America's air-trained population.

With respect to her own flying, however, Amelia Earhart staunchly refused to ask her husband's permission. It was her life's work, so precious to her that she was willing to risk her marriage rather than compromise herself. When George Palmer Putnam proposed to her in Los Angeles in the fall of 1930, she touched his arm lightly, nodded, and immediately jumped in her airplane and was gone for days. Much discussion between them ensued about Earhart's reluctance to marry. When Earhart awoke on her wedding day, February 7, 1931, at the Putnam family estate in Noank, Connecticut, she sent him a letter, the contents of which prophecied unions many decades later. She wrote:

> In our life together, I shall not hold you to any medieval code of faithfulness to me, nor shall I consider myself bound to you

> similarly. If we can be honest I think the differences which arise may best be avoided.
>
> Please let us not interfere with each other's work or play, nor let the world see private joys or disagreements. In this connection I may have to keep some place where I can go to be myself now and then, for I cannot guarantee to endure at all times the confinements of even an attractive cage.
>
> I must exact a cruel promise, and this is that you will let me go in a year if we find no happiness together.
>
> I will try to do my best in every way.

Their year's contract was successfully fulfilled, and their relationship continued as a fond and respectful, as well as a mutually beneficial one. That Putnam said "I do" in agreement with Earhart's revolutionary demands is proof that she had found an uncommon man indeed with whom to share her life. That she had the strength and personal honesty to insist upon them vaults her into a heroic role among women of her era which surpasses her pioneering profession.

Her forthright honesty extended to her position as an American heroine as well. For years, she kept a file in which she placed letters gushing with praise which she tabbed, "BUNK." Her insistence on credit simply for what she did, no more and no less, was as vehement as it was futile.

In the winter of 1934, Earhart set out after an air record which men pilots (among them Howard Hughes) had never successfully completed. A $10,000 prize was offered by a group of Hawaiian businessmen to the pilot who flew solo across the Pacific Ocean. Such a flight, conquering the 2,400-mile stretch between Honolulu and California, would be proof of an air link to the United States mainland.

Just before Christmas, Earhart, Putnam and her technical adviser for a Pacific flight, Paul Mantz, boarded the S.S. *Lurline* bound for Honolulu. Mantz had been an air corps pilot and was now a Hollywood stunt pilot, advising film makers on such films as *Hell's Angels* and *Men With Wings*. Considered a pilot's pilot, Mantz was an expert at fuel-conserving techniques and the design and use of navigation equipment. For the trip, Earhart's Lockheed Vega was strapped to the foredeck of the ship and covered with a tarpaulin.

"If I do not do a good job," Earhart told Putnam, "it will not be because the plane and motor are not excellent, nor because women cannot fly." She took total responsibility for anything that might happen on her long solo flight across the Pacific.

On January 12, 1935, Amelia Earhart became the first pilot, male or female, to fly solo to the continental United States from Hawaii. As she landed her Vega at Oakland Airport in California and stumbled out of the cockpit, having barely moved her knees for eighteen hours and fifteen minutes, she was rushed by a crowd of over ten thousand people. The same flight today takes only five hours, fifteen minutes, and passengers are frantically entertained with meals and movies the entire trip. Amelia Earhart, over four decades ago, suffered the eighteen hours alone in the cramped cockpit of her Vega, sitting with her maps on her knees, eating malted milk tablets and sipping hot chocolate from a thermos. But in contrast to her transatlantic flight three years earlier, Earhart's navigational information and equipment were so competent and the Pacific skies so accommodating that she let a broadcast of the San Francisco Symphony, tuned in on her air-to-ground radio, entertain her for the last part of her lonely trip. Instead of the fearsome storms of the North Atlantic, she was greeted by the friendly beams of searchlights blinking in greeting from ocean liners, alerted of her flight, along her route. Though still pioneering, aviation had indeed made progress.

After a long night's sleep, Earhart was inside the hangar at Oakland the next day conferring with mechanics and talking, somewhat preoccupied, with the reporters. The Vega was cordonned off, safe from curiosity seekers. Outside a crowd had gathered, held back by guards at the rope. A tall, sixteen-year-old girl with a camera pushed her way past the guards shouting, "Press!" and rushed in to join the journalists and photographers from the Associated Press, New York *Herald Tribune,* and other big-city dailies eager for a story on Earhart's flight. After several minutes of questioning, the lanky teenager with curly brown hair identified herself as Gene Shaffer from the University High School *Daily U-N-I* and asked Earhart for advice to high school students. The other reporters turned and looked at her aghast, but Earhart answered immediately. "Aviation is the career of the future," she said warmly. "You and your fellow students have the first chance in history to train yourselves and become a part of it."

Gene stood looking at the slender, five-foot-eight, 118-pound woman, built exactly like her, who had flown herself across two oceans. At that moment Gene felt the personal impact of her mother's favorite expression: "Hitch your chariot to a star."

After leaving Oakland, Earhart became the first pilot to fly nonstop to Mexico City and then from Mexico to New York. Gene

Shaffer would wait six years before fulfilling her desire to become a pilot. But her Earhart-like body, considered frail and underweight on the Army Air Corps cadet physical examination which had been unscientifically modified for feminine fliers, would almost disqualify her from war service as a valuable test pilot.

INVITED by the President of Purdue University, Edward V. Elliott, Amelia Earhart became the women's career counselor at the University's Indiana campus in the fall of 1935. Many of Purdue's 800 women students sat for hours listening to Earhart, in slacks, perched knee to chin on a table, as she expounded freely her hard-fought ideas on women's responsibilities to themselves, and the realities they faced when they took them. "Times are changing," she said one evening at an informal lecture, "and women need the critical stimulus of competition outside the home. A girl must nowadays believe completely in herself as an individual. She must realize at the outset that a woman must do the same job better than a man to get as much credit for it. She must be aware of the various discriminations, both legal and traditional, against women in the business world. . . . I'm inclined to say that if you want to try a certain job, try it. And if you should find that you are the first woman to feel an urge in that direction—what does it matter? Feel it and act on it just the same. It may turn out to be fun. And to be fun is the indispensable part of work."

Each senior that year had an hour's appointment with Earhart to discuss her future. In the spring of 1936, twenty-one-year-old Ruth Grimm sat down in Earhart's office for her interview.

"What is your major?" asked Earhart.

"Clothing design," said Ruth.

"Design is a good field," the aviator said. "It's not just clothes. It could be a can of sardines."

Ruth was so puzzled by Earhart's words that she called her father, a businessman in Indianapolis, to ask him what she meant. "She's right," her father said. Earhart's savvy view of Ruth's skill was just like a man's.

At the end of the school year, the annual women's banquet at Purdue featured Amelia Earhart. Unforgettable to a clothing design student was the dress Earhart wore that evening. It was sleeveless, midnight blue to the floor, and covered with silver stars as if the gods had sprinkled them on her. Around her neck was a cloudlike collar of white mink. But as Earhart spoke thrillingly about her trips alone across the world's great oceans, her enthusiasm for flying also made an indelible impression on Ruth Grimm.

After graduation from Purdue, Ruth Grimm married and read the newest books on designing the perfect marriage. The wife, one asserted, should attempt the same activities as her husband, for "togetherness." When Charley Trees announced he was taking up flying, his wife Ruth thought of Amelia Earhart and joined him in flying lessons. Within weeks, however, Charley's enthusiasm waned. But Ruth had found the most rewarding activity of her life. She returned to Earhart's urgings that women should find independence away from home. To show his agreement (and his modernity), Charley bought Ruth her own airplane.

Two years later, in 1940, when Charley joined the OSS and was sent to Burma, Ruth had an avocation (and a new baby) to keep her company. But in fulfillment of Earhart's dream, Ruth's avocation was soon to become a profession. Her flying skill—she had earned her commercial license—would take her, too, to war.

Purdue University provided Earhart with more than the opportunity to influence eager young minds. The Purdue Research Foundation inaugurated an Amelia Earhart Fund to buy her a "flying laboratory," the most sophisticatedly equipped, most advanced airplane available to a civilian, so she could conduct flying experiments. Unofficially, she began planning her next big flight, to take place in 1937—around the world at the equator.

WHEN Earhart first contemplated flying around the world, the feat had been performed only twice, both times by her friend Wiley Post, a taciturn, nearly invulnerable-looking Oklahoma barnstormer, who was part Indian and had only one eye. The other had been lost in an oil rig accident—he bought his first plane with the insurance money. In 1931, with a navigator, he flew a Lockheed Vega, like Earhart's, from Roosevelt Field, Long Island, around the world, 15,000 miles, via England, Germany, Siberia, Alaska and Canada. The flight took eight days. In 1933, he flew the same trip by himself. But in 1935, on his third attempt, this time with humorist-philosopher and fellow Oklahoman Will Rogers, Post crashed into a lagoon near Barrow, Alaska. Both men were killed. No one had yet attempted to fly the 28,000 miles around the world at the equator, the pioneer aviator's last challenge.

Shortly before her thirty-eighth birthday in July 1936, Earhart took delivery on a twin-engine Lockheed Electra, her Purdue "flying laboratory." As technical advisor for the Electra and its radio equipment, Earhart returned to her friend Paul Mantz, the dashing Hollywood stunt pilot who had assisted her on her Hawaii-to-Oakland solo flight in 1935. Mantz's job was not easy. Earhart was denied the electronics industry's most advanced radio equipment, which either belonged exclusively to the military or was prototype equipment manufacturers were unwilling to give to a woman for such a precarious venture. As to the equipment that was installed, so little was known or proven about its long-distance air-to-ground radio communications capabilities that no two technicians could agree on matters as simple as how long the aerial should be.

To circumnavigate the earth as close to the equator as possible, given fuel limitations of the Electra and existent runways long enough for taking off and landing, Earhart designed an itinerary. She, her navigators and technical advisor would take off from Oakland, California, and fly to Hawaii. Mantz would make the final check of the Electra there, and the three would head without him out into the central Pacific, aiming for Howland Island. After refueling there, they would fly to New Guinea, and the most perilous part of the trip would be behind them. From New Guinea, the next stop would be Port Darwin on the northern coast of Australia; then across Indonesia and India to Arabia; across Africa to Dakar, Senegal, the international airport used by airlines to make the jump across the South Atlantic, to the nearest point in Brazil, Natal. From Natal, Earhart would fly up the coast of South America, across the Caribbean to Puerto Rico, and then west toward Miami and home.

Preparation and arrangements for the flight were prodigious. Standard Oil Company agreed to provide fuel deposits at all of her stopping points; Pratt-Whitney, manufacturer of the Electra's two 550-horsepower Wasp Senior engines, shuttled spare parts along the route. Earhart was cleared with scores of embassies. She was even given letters in Arabic, prepared by two linguists in New York, in case she was downed in the Arabian desert and encountered hostile sheikhs. On the tiny Polynesian island of Howland, a United States possession in the middle of the Pacific—2,200 miles southwest of Hawaii, 2,500 miles northeast of Australia and over 4,000 miles from California—the Department of the Interior, by executive order of President Roosevelt, built her an airstrip. Howland Island would be her one refueling stop between Hawaii and Asia.

Female flying colleagues were fearful about Earhart's trip, especially the long over-water hops. On a hunch, Jacqueline Cochran, who, with her husband, millionaire Wall Street financier Floyd Odlum, helped fund Earhart's flight, made her test out her navigator, an experienced ship's captain. Earhart flew him out far into the Pacific from Los Angeles, circled a while, then let him navigate by the stars—all he would have on the actual Pacific flight—to bring them back. When they reached landfall, he had missed Los Angeles by 200 miles, hitting the coast far to the north toward San Francisco. Cruising speed in the Electra, around 150 miles per hour, demanded faster calculations than on a ship. To hit a tiny Pacific island, navigational error could not exceed a couple of miles.

Earhart selected another navigator to make the flight as well, veteran Pan American pilot and navigator Fred Noonan. Temporarily released by Pan American because of a drinking problem, he accepted this chance to redeem himself in his profession by navigating Earhart around the world.

Earhart's friend, Louise Thaden, a pilot with the Bureau of Air Commerce, made a special trip to California to try to dissuade Earhart from going. "You're tops, now," she argued, "and if you never do anything else you always will be." Earhart was committed. "I've worked hard and I deserve one fling during my lifetime," she said with determined levity. "If I should bop off, it will be doing the thing I've wanted to do most. The Man with the little black book has a date marked down for all of us—when our work here is finished."

LATE one March evening in 1937, twenty-one-year-old Carole Fillmore left her airplane wing half painted in the hangar at Oakland Airport and went out into the cool damp night in search of a cup of coffee. Suddenly a slight figure materialized out of the darkness and approached her. "Hello, are you a pilot?" said a soft voice. Carole looked at the woman, now lit by light reflected from the open hangar door. Her breath caught as she recognized the pretty freckled face and frank, shy smile of Amelia Earhart.

Carole knew from the newsreels that Earhart would be at the airport, for her departure point to Hawaii, that week.

"I'm just learning," Carole said, filled with modesty about her five years' flying experience when faced by the pioneer aviator. Earhart's eyes brightened. Carole told her she flew passengers for sightseeing trips around San Francisco Bay. At the end of each tour, she asked her passengers how they liked the flight. While doing so, she would whip off her flight helmet which let her hair fall to her shoulders. Her passengers were always dumbfounded to find they had just flown with a woman. Earhart laughed. "I understand what you mean," she said. Then she asked Carole if she would like to see her airplane. They agreed to meet early the next morning.

Forgetting all about her cup of coffee, Carole drifted back into the hangar in a daze. She felt a strange and powerful affinity for the lone figure, perhaps still out in the night, thinking of the import of the flight she was about to make.

On the morning of March 17, 1937, Carole Fillmore met Earhart by her gleaming metal Electra, its two Wasp Senior engines thrust proudly toward the sky. The wings were so high that Carole could walk under them without ducking. Earhart took her through the passenger compartment big enough for ten people but filled instead with a chart table and chair for the navigator and bulky pretransistor radio and navigation equipment. In the cockpit, Earhart explained the Sperry gyroscopic autopilot which flew the plane by electronic sensors, and the instrument panel. Carole was amazed at what she was used to calling the "dashboard." A twin-engine plane had two of everything. "It must be like flying two airplanes at once!" she said.

That afternoon, Carole watched Earhart and her crew lift off the Oakland runway. Her eyes followed the huge silver Electra until it disappeared beyond the Golden Gate Bridge. Seven years later, Carole herself would be flying airships that were much bigger, better equipped and twice as fast as the plane that had just vanished with her heroine, heading west.

Two DAYS LATER, Amelia Earhart's around-the-world flight aborted in Honolulu. On takeoff for Howland Island, a landing gear shock absorber collapsed (Earhart's diagnosis; others thought that the plane she had been flying for over nine months was too heavy for her to handle), causing one wing to crash into the ground. The Electra was crated and shipped back to the Lockheed Burbank, California, hangar for a complete overhaul. Earhart's flight was posptoned for three months.

Because of the change in seasons, the flight route was reversed. The ship captain navigator withdrew and Earhart and Noonan would make the flight alone, leaving from Miami, heading eastward to Puerto Rico, south to Natal, Brazil, across the Atlantic to Senegal, across 4,400 miles of Africa to Karachi (now part of Pakistan, then in India), on to Burma, down to Singapore, over Indonesia to Port Darwin, then to New Guinea, their departure point for the Pacific—Howland Island, Hawaii and finally home to Oakland.

Earhart agreed to file accounts of her trip along the way with Carl Allen, a reporter with the New York *Herald Tribune*. On June 1, 1937, Earhart and Noonan took off from Miami and the trip at last had begun.

For the next four weeks, Earhart wrote a flight log, like that of the *Friendship* nine years before. But this time, rather than Bill Stultz as her pilot, she let a Sperry gyroscope fly the Electra as she noted curiosities for the *Herald Tribune*. She took snapshots of burros in Fortaleza, Brazil, she wrote, and was amused by champagne bottles along the walkways of the International Airport in Natal. On arriving in Senegal, she mused about the beauty and dignity of native Africans, compared with their plight once the United States had transformed its Negroes. She lamented the total inaccuracy of maps of the African deserts, recounted camel rides in Karachi, tight squeezes through monsoons over the Bay of Bengal, pagodas like golden beacons in Rangoon and a day of exploring the craters of Java with a Dutch volcanologist and his two volcanic-gas-sniffing dogs. The jungles of New Guinea fascinated her, she wrote, as did the pidgin-English spoken by the natives and the "fairy-story sky country" over the New Guinea mountains. But she most wanted to be home by the Fourth of July and to have the vast stretch of the Pacific behind them. From Lae, New Guinea, she and Noonan would fly over the Solomon Islands and take aim on Howland Island.

To make the markerless over-water flight, they would have to navigate by the stars and sun. Clear skies were imperative. Their radio receiver would be tuned to a frequency on which a homing signal was broadcast on the hour and half-hour (a beeping letter "A" in Morse Code, or . –) from a U.S. Coast Guard cutter called the *Itasca,* moored off Howland Island. But this signal would not be picked up until two to three hundred miles from Howland. If one has tried to throw a paper wad across a windy field into a teacup, one can conceive of the difficulties of hitting a two-by-one-half-mile island in the middle of the Pacific Ocean from 2,550 miles away. On Howland was their only fuel until Hawaii.

Navigation would have to be accurate for another reason. In July 1937, Japan was expanding its military-enforced sphere of influence all over Southeast Asia and throughout the Pacific Islands. Just north of where Earhart would fly were the Marshalls and other island groups mandated to Japan by the League of Nations after World War I. Exact knowledge of the extent to which Japan was fortifying these islands had been impossible to obtain as the Japanese had forbidden almost all non-Japanese from setting foot on any mandated island, entering a port, or, should an aircraft have the range, flying over one. Japanese air space was known, even in 1937, to be swarming with Zeros.

Earhart and Noonan took off at 10:30 A.M. on July 2, from Lae, with 4,000 miles' worth of gasoline (up to 22 hours), one-and-a-half times what they needed to reach Howland. The Electra would be within reception range of the *Itasca*'s homing signal by around 7:00 A.M. the following day, which would be July 2 on Howland, across the International Date Line. In the radio room of the *Itasca* were Coast Guard radio operators, the ship's commander, several officers and crew members, a U.S. Department of the Interior field representative and Associated Press and United Press International reporters, eagerly awaiting Earhart's first transmission.

At 5:20 P.M., Lae received word from Earhart that she was 798 miles out, the weather was clear, and they were on course. Not until over nine hours later, at 2:45 A.M., did the *Itasca* radio control room hear its first transmission from the Electra. Earhart's voice reported cloudy and overcast conditions, but her words were barely intelligible through static. A radio check verified the clarity of the *Itasca*'s equipment. Earhart's radio was evidently malfunctioning.

As July 2 dawned a partly sunny day on Howland Island, the increasingly anxious occupants of the *Itasca*'s radio room heard sporadic, garbled transmissions from the Electra indicating Earhart was still flying in a storm. She had answered none of their pleas, mounting in urgency, for a position report, and gave no indication that she could pick up the homing signal. At the hour when she should have been directly over Howland, Earhart broadcast, her voice now strained, that she thought she was near and had descended to 1,000 feet searching for the island. Given local weather conditions, she should have been able to see it from miles away. At 7:58 A.M., she radioed that she was circling, still trying to locate Howland and the *Itasca*. By now the radio operators were frantic. The homing signal was beamed continuously. Transmissions from the Electra were too short or too weak for them to get a bearing on its whereabouts, however. It now seemed she was receiving none of the *Itasca*'s radio transmissions.

At 8:45, the *Itasca* radio room heard Earhart's tense voice crackle over the speaker: "We are on a line of position 157—337. Will repeat this message on 6210 kilocycles. Wait, listening on 6210 kilocycles. We are running north and south. . . ."

These were the last words heard from Amelia Earhart, America's most admired and honored heroine of the age. By midmorning, word had reached the White House in Washington that she had not arrived at Howland. President Roosevelt ordered the largest search in United States Naval history for America's First Lady of the Air, mobilizing a battleship, four destroyers, a minesweeper and the aircraft carrier *Lexington,* which immediately set out from California with fifty-seven airplanes on deck, for the central Pacific. For two weeks, ships and airplanes searched 250,000 square miles of the Pacific. Across the United States, day followed day with headlines, as ham operators scanned for signals and prayer services were held. As July 1937 came to a close, not a trace of the Electra or its beloved pilot had been found. The *Lexington,* bereft of fuel and hope, sailed under the Golden Gate Bridge, its flags at half-mast.

Many Americans could not accept that their heroine might simply vanish. Seven years later, in 1944, as island after island was taken from the Japanese during World War II, the search for Amelia Earhart continued, with Pentagon sanction. Rumors persisted that the Electra had been downed by Japanese Zeros in the course of spying over the mandated islands, and that Earhart and Noonan had been tortured and executed. Other theories held that she was still alive, living under an alias in relieved anonymity. But experienced fliers accepted a more simple end, one considered inevitable at any moment for them all, to their tousle-haired colleague and friend. "I feel now, as I did then," wrote Ruth Nichols, her fellow 1933 Bendix Race competitor, in 1958, "that Amelia flew on across the trackless Pacific until the last drop of fuel was gone and then sank quickly and cleanly into the deep blue sea."

IV

Jacqueline Cochran: On a Track to the Stars

In October 1934, an air race unprecedented in its distance—halfway around the world—was held between Suffolk, England, and Melbourne, Australia. For aviation historians, it was the most spectacular air race ever, for it began a new era in air transportation. Airplane manufacturers renowned today were already testing prototypes on the 10,000-mile race. Douglas built a DC-2, Boeing a 247, and they placed second and third in the race, to establish reputations that vaulted them to the forefront of commercial aviation.

Over sixty thousand aviation enthusiasts jammed the Royal Air Force base at Mildenhall, Suffolk, to watch the twenty hardy pilots take off for the other side of the globe. To everyone's amazement, two of the aviators were women. One was twenty-seven-year-old Amy Johnson Mollison, an Empire-wide celebrity.

The other was an American named Jacqueline Cochran. Tongues were wagging in the small world of aviation about this fledgling woman pilot. According to rumor, she had completed a rare and coveted course in "blind flying"—flying not by reference to check points on the ground, but by compass, airspeed and the second hand of her watch. Her soft drawl betrayed a rural Southern childhood. But when she talked, it was clear that she was no simple Southern belle. Behind the large brown eyes and soft, vulnerable mouth of the pretty blonde woman was a mind with a businesslike toughness. She had been a successful beautician at the fashionable Antoine's salon at Saks Fifth Avenue in New York.

Otherwise, little was known about Jacqueline Cochran. For air race enthusiasts, the England-Australia competition was her debut. Cochran and her experimental Gee-Bee racing plane would be forced down in Bucharest, Romania, with a sputtering engine and defective landing flaps.

In 1935, back in the United States, Jacqueline Cochran reappeared in Los Angeles at the starting line of the Bendix Transcontinental Air Race. The Bendix Race, in which many speed records were set, was the high point of the National Air Races, an annual, weekend-long series of speed trials and exhibition flying which in the 1930s merited front-page headlines in all of America's major newspapers. Amelia Earhart had been the first woman to compete in the Bendix Race in 1933. For the 1934 race, women were banned because earlier in the year a woman racer had been killed in a Chicago air meet. But in 1935, Earhart prevailed on air race officials to let women enter again. She flew her Vega to Los Angeles, and was joined by a glamorous newcomer, Jacqueline Cochran, who would fly a Northrop Gamma. Before the race began, Earhart and Cochran had time to get acquainted.

As the starters took off in the middle of the night on May 20, 1935, Earhart got away without difficulty. But by the time Jacqueline Cochran was next on the ramp, the sky seemed to drop like a tarpaulin, as a thick sea fog descended over the airport. Cochran watched anxiously as the pilot just before her roared down the runway and faded from view. Seconds later, there was a muffled explosion and the fog lit up eerily in the distance. Almost involuntarily, because of the training she had had as a nurse in her adolescence, Cochran got in her car and sped down the runway behind the fire truck. But when the fire was extinguished, the pilot was beyond help. He was dead.

Even before the fog, the manufacturers of her airplane and its engine had tried to persuade Cochran not to enter the race. Though the companies had felt no qualms about selling their Gamma model to Trans World Airlines to carry mail across the country, they were not so eager to see it tested in the rigors of a transcontinental air race demanding high performance and, if entered by a woman pilot, inviting high visibility. But Cochran told them she was determined at least to start the race.

As a wrecker truck removed the demolished plane from the runway, Cochran returned to her Gamma. Bystanders spoke in muted shock about the crash. When she overheard a government aviation official voice his opinion that taking off in such a fog was suicide, Cochran ran behind the hangar and threw up. Then she went to the nearest telephone and dialed her fiancé, Floyd Odlum, in New York. "What should I do?" she asked. "There's a fine line between a course of action determined by logic and one dictated by great emotional urge," Odlum said. "No one can draw that line for another. It simmers down to a philosophy of life." Cochran hung up and climbed into her airplane.

At 3:00 A.M. the Gamma lifted off the runway, but barely. With its heavy fuel load, it flew so low that the radio antenna hanging below the belly was torn off by the outer fence of the field. To gain enough altitude to clear the 7,000-foot mountains to the east, Cochran had to spiral by compass up into the fog over the Pacific. While doing so, her engine overheated with the strain of its load.

By dawn she was into clear skies and heading toward the Grand Canyon, distinct in the distance. Suddenly the tail of the Gamma began to vibrate violently, as if to shake the airplane apart. Cochran's eyes darted to the left and right, but there was no landing field for hundreds of miles, only peaks and empty desert. The engine was still dangerously overheated. "At least I started the race," she thought, "and it was a blind takeoff at that. I'm going back to Los Angeles."

Safely back at the Bendix Race starting line, Cochran told reporters she had been "too exhausted to go on." She refused to blame her failure to finish the race on a machine or its manufacturer. (In the Bendix Race the following year, a Gamma exploded in midair, catapulting its pilot, Joe Jacobson, into the sky; at the last minute before taking off, fortunately, he had slipped on a parachute.) When the race finished in Cleveland, Ohio, hours later, Amelia Earhart had placed fifth. For Jacqueline Cochran, success and fame were still two years away.

FIVE MONTHS after Amelia Earhart vanished without a trace, about two hundred people huddled near a wooden speaker's platform by the administration building of Floyd Bennett Field in Brooklyn. The skies were a dismal November gray, and the crowd tightened their winter coats against the freezing west wind which whipped through two American flags mounted at each end of the platform. The Women's National Aeronautical Association was paying tribute to Amelia Earhart, but the speaker, Jacqueline Cochran, conveyed the feelings of everyone that, whether or not Earhart still existed somewhere in the flesh, her spirit was very much alive.

"If her last flight was into eternity, one can mourn her loss but not regret her effort," Cochran said somberly. "Amelia did not lose, for her last flight was endless. Like in a relay race of progress she had merely placed the torch in the hands of others to carry on to the next goal and from there on and on forever."

Cochran's voice was clear but her eyes glistened as she spoke. Another American flag, a small one of silk, lay folded among her special memorabilia. Amelia Earhart had given it to her in June, just before she left for Miami on her last flight, which Cochran and Floyd Odlum, now her husband, had helped to finance. Though neither woman knew of its import at the time, the small silk flag was to symbolize a passing of the torch.

OVER Labor Day weekend, 1937, a month after Earhart was given up as lost, Jacqueline Cochran entered the Bendix Race as the event's only woman, and with that race she began to win. Not only did she gain the $2,500 Bendix women's purse, but she finished the race in third place, ahead of several of America's toughest men competitors.

There could never be another Amelia Earhart, but aviation no longer needed a pioneer to champion women's right to the airways. World politics was focusing America's attention away from exploratory flights to the far reaches of the earth, and onto the menacing fascist regimes of Europe and Asia. Five days after Earhart failed to arrive at Howland Island in the Pacific, Japan launched a full-scale invasion of Peking and other major Chinese cities, beginning the Sino-Japanese War, and bringing a new foreboding to the 1936 Anti-Comintern Pact between Japan and Germany, and thus to the Rome-Berlin Axis. Every year, since 1935, Congress had passed Neutrality Acts. America had wrapped its wings around itself in determined isolationism.

Reflections in the press on Earhart's last flight revealed a prevailing change in the country's view of aviation, as well. The DC–2 of the England-Australia air race had developed into the DC–3, the first stable and safe modern airliner, and with it and similar models, American manufacturers were beginning to supply the world.

The airplane was coming of age, and the new First Lady of the Air could no longer be a romantic vagabond—she had to be a competitor, a glamorous, gutsy and influencial participant in the growth of the industry of the future. In Jacqueline Cochran, the nation had found her.

Cochran smashed one speed record after another, culminating in a flight on December 4, 1937, when she raced from New York to Miami in four hours, twelve minutes to set a national record, supplanting that of the dashing racing pilot and millionaire Howard Hughes. For the flight, she had calculated her fuel supply so carefully, to eliminate every ounce of excess weight, that as she touched down onto the runway in Miami, her engine coughed and died, absolutely dry. She had to be towed to the hangar, as cars full of enthusiastic reporters and fans rode along side as her retinue.

The year 1938 began with a rain of awards for America's new air heroine. But though Cochran appeared to have reached the top of the aviation world, before 1938 was through she would achieve a pinnacle, wholly unexpected by her aviation colleagues and unimaginable for a woman in the mind of her enthusiastic public.

THOUGH Cochran had placed third in the 1937 Bendix Race, her flying time from Los Angeles had been a full two and one half hours longer than the air race winner's, Frank W. Fuller. Fuller's plane was a new military-type pursuit plane, the P–35, designed by a Russian World War I Ace, emigrated to Long Island, named Alexander de Seversky. Its engine was an unheard of 1,000 horsepower, and the sleek, all-metal, low-wing body made the other airplanes in the race look like antiques.

For many months, Seversky had tried to sell his P–35 model to the Army Air Corps, without success. Pursuit planes, according to prevailing Air Corps strategy, were for short forays to defend an air base from attack. With his P–35, Seversky was offering a fast, long-range pursuit plane which could accompany and protect bombers on raids far from base.

As words were getting him nowhere, Seversky decided to make his case with performance. Over the next few months, as Seversky watched Jacqueline Cochran emerge as the new American air heroine, he got an idea. Frank Fuller could enter his P–35 in the 1938 Bendix Race. But if a woman also flew one, and placed third, as Cochran had done in 1937, the Air Corps would have to take notice. Furthermore, she could test a fuel system innovation he had been developing. In an otherwise standard 1,200 horsepower P–35, he would build extra fuel tanks into the wings so that she might fly the entire race nonstop; the stretch from Los Angeles to Cleveland had never been done in one flight. When he telephoned Cochran in New York, all of his skillful arguments why she should take the risk were unnecessary. She immediately said, "Of course!"

The 1938 Bendix Race offered the largest purse in history—$24,000 would be distributed to the winners. Out of twenty-one pilots who attacked the pretrial measured course over which they had to maintain 225 miles per hour, ten qualified to enter the race: nine men and Jacqueline Cochran.

By midnight, September 2–3, 1938, parked cars jammed streets for blocks around Burbank Airport as the Bendix Race began. At 3:13 A.M. forty thousand spectators watched in tense silence as the third racer roared down the runway in her silver P–35, its rounded, clear canopy top shining in the field lights. After she broke ground, the plane climbed so steeply that the crowd cheered. In seconds she had disappeared into the darkness.

Shortly after dawn, another crowd gathered at Cleveland Municipal Airport for a day of aviation events that would culminate in the arrival of the Bendix racers from California. By midafternoon all eyes peered westward. Rumors were circulating that weather conditions between Cleveland and Los Angeles had turned stormy, and the crowd fell into a restless lull of anticipation for the arrival of the first Bendix racer.

At 2:23, a silver plane swooped down across the finish line. The crowd cheered. The timers clicked their stopwatches, and over the loudspeaker came the announcement that Miss Jacqueline Cochran, in her gleaming P–35, had just won the Bendix Race. She had flown the 2,042 miles from Los Angeles in eight hours, ten minutes and thirty-one seconds—nonstop.

As the P–35 rolled to a halt, reporters ran onto the ramp. Cochran kept them waving their notebooks and press cameras while she put on fresh face powder and lipstick in the cockpit, then slid back the canopy and emerged smiling radiantly. "Where's my husband?" she asked, as her first words to the press. Then she asked if anyone had a cigarette. "I've smoked an oxygen pipe all the way across the country!" The reporters laughed and barraged her with questions about the storms along the route. "Yes, it was pretty tough," Cochran said modestly. "I saw the ground maybe for half an hour the whole trip. But my airplane functioned perfectly."

The reporters surveyed the sleek P–35 appreciatively and took down her words. They also recorded her as the "first woman to conquer the problems of blind flying." Then a car drove up and whisked Cochran away to find her husband, Floyd Odlum.

Cochran did not tell the press that she had landed with less than three gallons of gasoline, only minutes' worth, left in her tanks. Forced to climb up to 22,000 feet above the storms, she had encountered headwinds and the trip had taken twenty minutes longer than she had planned. Nevertheless, the Bendix Race had at last been flown nonstop. Frank Fuller would not arrive to place second in his P–35 for another hour. Without the extra wing tanks, he was forced down in Wichita to refuel.

Another detail reporters did not know at that time was that the experimental wing tanks in Cochran's P–35 had caused her some anxious moments. After she climbed to 20,000 feet heading toward the Rocky Mountains, her engine had sputtered and quit. When switching gasoline tanks did not start it again, she tried rocking her wings. The engine caught. The right wing tank, she discovered, was not feeding its fuel to the engine evenly. By elevating the wing periodically throughout the trip, she was able to drain out gas by gravity. After the race, a wad of paper was found inside the wing tank by the gas drainage valve. Unlike the streamlined manufacturing processes of Detroit, the airplane assembly line was still far from meticulous. Cochran dismissed suggestions of sabotage.

The beaming blonde Bendix Race winner was on the ground in Cleveland just long enough to see her husband and make a short speech to the stands. Then she took off for Bendix Field, New Jersey, to set a new women's cross-country record of ten hours, thirteen minutes. And at seven o'clock Saturday evening, Cochran boarded a commercial airliner and returned to Cleveland in time to make a grand entrance as the belle of the air race ball.

Alexander de Seversky, designer of the victorious P–35, was also celebrating. His idea to shake up the Air Corps by having a woman fly his plane worked so well that for several days following the race, certain officers claimed that Jacqueline Cochran had not flown the race herself, but had been replaced under cover of darkness by a male pilot. But when the rumors had run their preposterous course, Seversky started taking orders. The P–35 developed into the P–47 Thunderbolt fighter, the most effective long-range pursuit escort of the African and European theaters during World War II. Seversky would demonstrate his gratitude to Jacqueline Cochran six years later, when at an important moment, he would affiliate her with his entire production of P–47 Thunderbolts.

In March 1939, Jacqueline Cochran was awarded her second Harmon Trophy, the highest award given to any aviator in America. But as America's top woman aviator, she refused to bask in the gleam of the rapidly filling trophy case in her Manhattan apartment. The day before the Harmon award was announced, she had broken a women's altitude record by climbing to 33,000 feet above sea level. As she did for all her racing flights, she had trained for the attempt like an athlete, with exercises, special high-protein diets and many pep talks, both from herself and her supportive husband. Such preparation was warranted. For each thousand feet she climbed, the temperature dropped 3.5 degrees, a crucial pilots' statistic that she dreaded on this flight. Her 600-horsepower "stagger-wing" Beechcraft, in spite of its racy sounding name, was in actuality no more than a fabric-covered bi-plane—with no heat.

She was only at 12,000 feet when her thermometer hit zero. By 20,000 feet, the temperature in the cockpit was minus thirty. As she climbed through 30,000 feet, it was sixty degrees below zero. Nor was the cockpit pressurized. She withstood the cold as long as she could, and had decided to begin her descent, when she felt a sharp throb in her temple, which she learned later had been a blood vessel rupturing in her sinus. She throttled back and pointed the nose of the plane down gently toward the earth. But as she re-entered the richer atmosphere she became totally disoriented. She had no oxygen mask, and the tube on which she had been sucking provided an uneven flow at best. She maintained her glide path but felt so light-headed that it took her an hour, circling Long Island, before she could get her bearings and find the airfield. When the telephone rang with the news she had won the Harmon Trophy the following day, she winced from a splitting headache which lasted for several days.

In June, she met Eleanor Roosevelt, as the First Lady again handed her the impressive bronze Harmon Trophy in a ceremony in Washington, and honored her for the Bendix Race victory and her women's altitude record. Over the next two months, Cochran broke two women's and two national speed records over measured courses, and one intercity record—between Burbank and San Francisco. Before test flying became totally war related, Cochran offered herself as a test pilot—to Seversky, the Mayo Clinic and anyone else who asked. On each of her record-breaking flights, sustained only with a half-filled bottle of Coca-Cola (a full one would explode at high altitudes) and a fistful of lollypops for "dry mouth," she tested new types of oxygen masks, engine superchargers, sparkplugs, airplane fuel and wing designs which would appear in the airplanes soon to become America's air arsenal.

Why did she take such risks on each flight? she was repeatedly asked. The answer was simple. "I might have been born in a hovel," she said, "but I determined to travel with the wind and the stars."

Frequent press photographs of the brown-eyed beauty applying lipstick in the cockpit belied the Dickensian childhood of Jacqueline

Cochran, during which she scrambled shoeless for sustenance, as a foster child, in the lumber milltowns of northern Florida. She literally did not know when she was born, or to whom.

At eight years old, little Jackie Cochran followed her foster family to Columbus, Georgia, where jobs were plentiful in local cotton mills. Her family rented a house with a bathroom, Jackie's first. Though only eight, she also got a job, working the twelve-hour night shift in a cotton mill, and the next year was put in charge of fifteen children in the fabric inspection room. Child labor cut off her formal education after only two years. But befriended by a local school teacher, Jackie learned to read.

From the cotton mill, Jackie was hired by the owner of a beauty shop to do odd jobs. By the time she was thirteen, she was cutting and styling as a full-fledged beauty operator. In the early 1920s, she became one of the first to learn how to give the newly developed permanent wave. The torturous process took all day.

Independent at last from her downtrodden foster family, which she continued to support for the rest of their lives, Cochran left Georgia to demonstrate permanent waves in beauty shops in Alabama and Florida. One of her customers, a woman judge in Mobile, persuaded the bright young beautician to attend nursing school.

After three years' study at a local hospital, she accepted a job with a country doctor in Bonifay, Florida, a milltown like those in which she was raised. Several grueling weeks later, Cochran reached a moment of truth. When she delivered a baby, by corncob and oil "mojo" lamp in a shack, the family did not even have a piece of cloth to wrap around it. "I had neither the strength nor the money to do the smallest fraction of what had to be done for these people," she recalled in her 1954 autobiography, *The Stars at Noon*, "and I determined that, if I was going to do anything for myself or others, I had to get away and make money. It seemed to me that I was adding as much of the happiness of others by making many of them beautiful as by making a few of them well—and I would certainly be doing a lot more for myself."

Though only in her early twenties, Cochran was soon hired by Antoine's at Saks Fifth Avenue and commuted between his salons in New York and Miami, while amassing commissions and meeting the cream of society's fashionable crop. The beauty salon, however, became as confining as the shacks where she spent her childhood. Dining at a club in Miami in 1932, Cochran was seated next to a Wall Street millionaire financier named Floyd Bostwick Odlum, to whom she expounded on her dreams of escaping the routine of a beauty operator by establishing her own cosmetics business. Odlum was sympathetic. The son of a poor Methodist parson, he had done his share of manual labor to work his way through college and law school. After a stint as a utilities lawyer, he had set up his own company, the Atlas Utilities and Investors Company, Ltd., which by 1928, when Odlum was only thirty-six, had total assets of $6 million. The canny businessman pondered Cochran's prospects. To cover enough territory, with the Depression on and sales competition so very keen, he warned his attractive companion, she would almost need wings.

In the summer of 1932, while crowds cheered Amelia Earhart for her courageous solo flight across the Atlantic, Jacqueline Cochran took her vacation from Antoine's on Long Island—at the Roosevelt Field airport. In two and a half weeks, she earned her pilot's license. A few months later, she left Antoine's for good and went to San Diego where she persuaded a Navy flight instructor to give her the equivalent of the entire U.S. Navy flight training course. As she was to write in her autobiography, "At that moment, when I paid for my first lesson, a beauty operator ceased to exist and an aviator was born."

AFTER his momentous, if inadvertent piece of business advice, Floyd Odlum became an ardent supporter, and soon suitor of Jacqueline Cochran. With Odlum's help, Jacqueline Cochran Cosmetics, Inc., was born.

Floyd Odlum and Jacqueline Cochran were married on May 10, 1936. By this time, the sandy-haired, slender and fast-moving Wall Street investment genius, now president of the Atlas Corporation, was being compared to J. P. Morgan in the extent of his control of American industry. Specializing in large-scale financing, Odlum had bought and turned around several floundering behemoths—the Greyhound Bus Lines, Madison Square Garden and Bonwit Teller—as well as various banks, real estate companies, motion picture producers and aviation concerns, among them the Curtiss-Wright Corporation. He would soon take on the Convair Aircraft Company. Among his many interests, aviation was always an enthusiasm for Odlum. He also considered himself a New Dealer, and contributed regularly to the Presidential campaigns of Franklin Delano Roosevelt.

With marriage, Cochran set about creating the homes she never had as a child. The Odlums bought a country estate in Stamford, Connecticut, a ranch outside of Palm Springs, California, and a large

Manhattan apartment overlooking the East River. Cochran's cosmetics company offices moved into New York's splendid new Rockefeller Center complex on Fifth Avenue. Their shared passion for aviation pervaded their home as well as their professional lives. On the walls of the Odlum apartment's entrance foyer were murals depicting pioneer attempts at flight. Suspended from the ceiling was a small chandelier fashioned after an observation balloon. Along the entrance hallway was another mural, picturing airplanes which Cochran and their aviator friends had flown.

But on a street nearby was evidence of another, equally deep-seated commitment. Soon after she was married, Cochran bought a building and established what was soon considered a model orphanage.

Relentless with herself as a competitor in the air, Jacqueline Cochran was just as vehement about her womanhood. For Cochran women were special, with distinctive abilities that she felt made them excellent pilots. "Women have the advantage of keen sensitivity and intuition," she said. "They often show more patience than men. They are accurate in calculations and have a fine sense of detail which nicely equip them for difficult flying."

Having lived by her wits all her life, she was uniquely qualified to achieve what she did for women in aviation. But, her "unfeminine" behavior caused her at times to be unpopular, even openly thwarted. Whatever the odds faced by the self-made men of Jacqueline Cochran's generation, hers as a woman were twice as harrowing, and fighting them took twice the tough self-possession. She was reaching the top of the aviation profession just when it was becoming massively and institutionally male—on the threshold of war.

By September 1939, there was no longer any doubt about the Axis powers' intentions to wage war in Europe. In March, Hitler occupied Czechoslovakia and helped Franco's forces triumph over the Republicans in Spain. In April, Mussolini invaded Albania. On September 1, German *panzer* divisions thundered into Poland, and the world was again at war. With the invasion of Poland, Hitler also gave the world notice that its next war would be fought substantially in the air. The invading ground troops were supported by intensive aerial bombing—known as the *blitzkrieg*. After a month of superhuman resistance, Warsaw at last capitulated on September 27, and the world wept at the news.

The next day, Jacqueline Cochran wrote to Eleanor Roosevelt. Given the drastic events occurring in Europe, it was none too soon to begin contemplating the idea that women pilots could fly in non-combat roles, and thus release men for war duty, she pointed out. But, Cochran continued, the effective use of women pilots "requires organization in advance." Implied in the letter was Cochran's willingness to devote her own time to planning that organization.

Eleanor Roosevelt thanked Cochran for her suggestion, and assured Cochran of her ultimate faith in women's potential contribution in the air and in all war support functions. But both women no doubt knew the decision would not be theirs. During World War I, pioneer woman aviator Ruth Law had offered her flying skills to the Army Signal Corps, to which the fledgling air corps was attached. Instead, the government sent her around the country to recruit men pilots and raise money for war bonds. In 1930, a proposal to the chief of the Army Air Corps to use women pilots in a military capacity was resoundingly rejected as "utterly unfeasible." Women were "too high strung for war-time flying." There was no guarantee that the achievements made by women like Cochran had modified that view.

Nevertheless, as a prominent public figure, Cochran spoke out on the potential role women pilots could play should America go to war.

As Cochran spoke, battles were raging in Norway, the Netherlands, Belgium and Luxembourg. On June 4, 1940, the British were defeated in France and from Dunkirk retreated back across the Channel; on June 14, the German armies marched into Paris. By mid-July, the Battle of Britain had begun, and it would wreak destruction for five long months.

In spite of the horror of a Europe besieged, America remained ostensibly neutral. But on September 15, 1940, Congress voted compulsory military service—the first peacetime draft in history—and called for 800,000 men to be conscripted within a year. The year would be up, and many more men called, three months before Pearl Harbor.

The day Congress instituted the draft, Jacqueline Cochran addressed a meeting of the Ninety-Nines, the women pilots organization, being held at the New York World's Fair. "There should be an organized women's air corps auxiliary to the other air forces in the government—controlled, supervised and supported by our government," she asserted. But not until the spring of 1941 was she given her first opportunity to participate in the war effort.

In late March, 1941, Cochran attended an aviation awards ceremony at the White House. Afterwards, she had lunch with General H. H. "Hap" Arnold, chief of the U.S. Army Air Corps, and Clayton Knight, who was directing American recruiting efforts for the British

Air Transport Auxiliary. On March 11, 1941, Congress had passed the Lend-Lease Act. Among the first American-made war matériel offered to war-ravaged Britain was a fleet of twin-engine Lockheed Hudson bombers, for use by the RAF in coastal defense and anti-submarine protection for convoys in the Atlantic. American manufacturers would get the airplanes to Canada; it was the responsibility of the British Air Transport Auxiliary to ship or fly the planes to Britain. With every possible qualified pilot involved in combat, the British ATA was desperate for recruits from neutral nations. It had even enlisted some of Britain's women pilots to ferry planes around the British Isles. But the transoceanic hope in bombers, Knight told his luncheon companions, were proving to be fearsome deterrents to his recruitment efforts.

"What can I do to help?" Cochran asked. General Arnold turned to her and made a suggestion. "Why don't you do some of the ferrying? They need pilots, and as a woman flying bombers you can publicize that need."

Knight was delighted. But when Cochran contacted ATA headquarters in Montreal, her offer met with a chilly reception. "We'll call you," she was told.

After several weeks, no call was forthcoming from Montreal. Cochran decided she had to use her aviation contacts. Lord Beaverbrook, the British newspaper publisher, and a friend of hers and Floyd Odlum's, had recently been appointed Churchill's minister of procurement (formerly of aircraft production) and she wrote him about what she wanted to do. The second week in June, she was invited to Montreal for a flight test.

Testing procedures for the 1938 Bendix Race winner and holder of seventeen aviation records proved to be more a test of will than of flying ability. Over the course of three full days, she was made to execute sixty takeoffs and landings, ten of which were "blind" from the back seat of an advanced trainer called a Harvard. A two-hour flight test is enough to exhaust most pilots, but Cochran was too determined to tire. Only one thing bothered her. The plane had no toe brakes, and she was forced to use a hand brake to the right of her seat in order to slow the three-ton plane on landing and to taxi. Late afternoon the third day, the chief pilot announced that her test was over. She breathed a sigh of relief, and asked if he would do her a favor and operate the hand brake while she taxied to the hangar. After three days, she joked, her arm was pretty sore.

When the chief pilot made his report, he stated that Cochran was qualified to fly the Hudson bomber, but he recommended against permitting her to fly one for the ATA. He felt there was a possible physical disability to handle the brakes in an emergency.

ATA headquarters overruled his objections as petty, and gave the order that Jacqueline Cochran ferry a Lockheed Hudson from Montreal to Prestwick, Scotland. She was assigned a copilot/navigator and radio operator. Immediately, however, protests erupted among the ATA pilots in Montreal. They called a mass meeting and threatened to strike. If the Germans shot her down, they argued, the ATA would be blamed for the death of America's most famous woman aviator. Others complained that if a woman flew a bomber across the Atlantic—and an unpaid volunteer at that—their jobs automatically were belittled.

The ATA high command compromised with its pilots. Cochran would be pilot in command across the Atlantic, but her copilot/navigator would take off and land.

On June 17, Cochran and her crew walked out to their Lockheed Hudson. All three were somewhat disconcerted by its condition. As Cochran checked the airplane, her responsibility as first pilot, she found the window in the pilot's cabin had been smashed. Once inside the fuselage, she noticed the life raft was missing and the wrench used to activate the oxygen system had been removed. While the window was being repaired, Cochran bought a wrench from a passing mechanic.

Finally, late in the afternoon, the Lockheed Hudson took off from Montreal, and Cochran took over the control wheel. Twelve long hours later, Cochran had directed the bomber on course straight across the Atlantic. She set up her final approach into the landing field of ATA headquarters, Prestwick, Scotland, then relinquished control back to her navigator for their landing.

Once again, the British press greeted an American woman pilot who had flown the Atlantic, this time in a bomber. But after proving a woman could fly a much needed bomber across the ocean, Cochran had other business in Britain. She went to London, and met with Captain Pauline Gower, chief of the women fliers of the Air Transport Auxiliary, before heading back to the United States of America.

On her return, July 1, 1941, in a news interview at her home, she spoke freely about the results of her trip. She went, she said, "Because I wanted to check up on what those girls were doing, and to see how we could organize a similar group in this country."

When the press finally left the apartment, the telephone rang.

She was invited to luncheon the next day at Hyde Park with President and Mrs. Roosevelt. A police escort would pick her up.

After lunch, Cochran spent two hours with President Roosevelt. She left Hyde Park with a note of introduction to Assistant Secretary of War for Air, Robert A. Lovett. In the note, the President gave instructions that Jacqueline Cochran was to research a plan for an organization of women pilots to serve with the United States Army Air Corps.

V

A Sky Full of Women

THE WOMEN who forty years ago not only yearned to be airborne but actually climbed into the cockpit had more in common with their pioneering predecessors than with the majority of their contemporaries. Flying was still an inordinate commitment around which they had to design their lives. In 1939, only slightly over one fourth of America's women earned their own money—in fact, the same number of women worked in 1940 as in 1910. Full-time salaries averaged from $1,100 to $1,200 a year for clerical staff, but women's average annual income overall was just above $850. A pilot's license cost between $500 and $750, and sometimes more. Most young women, of course, married, and if they did not find air-minded husbands to pay flying bills, they simply had little chance. Having a job of their own was frowned upon—before Pearl Harbor; only 15 percent of America's wives worked, and to put the Missus in the air would have cost half of the median national income for a 1940 family.

Yet the attractions of flight were so powerful that some women (and men) in prewar America were willing to redefine their lives and defy the risks. The reasons why lie in the dual nature of "flight" itself—the word in our language implies both to fly and to flee.

In flying there always has been the thrill of accomplishing a more than human activity. Preserved by history are stories of "bird-men" kings who, strapped to winglike contraptions, jumped from the towers of Trinavantum (now called London) and the minarets of Constantinople. In their desire to fly, human beings have sought a sense of extraordinary power, of elevated perspective and a special status above.

When flying actually became possible, a new attraction was added —the satisfaction of mastery. By attaining a knowledge of aerodynamic principles and the workings of a machine, one could achieve a miracle, flight. Every takeoff was a personal triumph, as well as a conquest of one of the most primal human fears, falling.

In flying, moreover, joyous, heightened sensations were available to the flier in control of his or her machine. To fly was to experience a unique and expanded visual and kinetic awareness that permanently changed one's sense of space and one's concept of what the world looked like.

But flight also appealed to many because it took them someplace—away. Not only an aesthetic experience, flight was an expression of independence and free will, a triumph over the eternal static hold of gravity. The ability to fly provided an opportunity not only to excel, but also to escape from the labyrinth of routine existence.

In the psychology of dreams, a realm explored concurrently and with equal fervor as the air, psychoanalysts have inevitably found sexual implications in flight, as well as in the airplane. Sigmund Freud, in Volume XV of his *Introductory Lectures on Psychoanalysis*, proferred:

> The remarkable characteristic of the male organ, which enables it to rise up in defiance of the laws of gravity, one of the phenomena of erection, leads to its being represented symbolically by *balloons, flying machines* and most recently by *zeppelin air ships*. But dreams can symbolize erection in yet another, far more expressive manner. They can treat the sexual organ as the essence of the dreamer's whole person and make himself *fly*.

Freud conceded that women, too, may have dreams of flight, but their symbolism is different for the female dreamer:

> Remember that our dreams aim at being the fulfillments of wishes and that the wish to be a man is found so frequently, consciously or unconsciously, in women. Nor will anyone with a knowledge of anatomy be bewildered by the fact that it is possible for women to realize this wish through the same sensations as men.

Worthy of note is Freud's dream imagery for female sexual references:

> The female genitals are symbolically represented by all such objects as share their characteristic of enclosing a hollow space which can take something into itself; by *pits*, cavities and hollows, for instance. . . . [Author's italics.]

Whatever the reasons why women climbed into the cockpit, their desire did have to be strong. For the vast majority of young women who had done nothing on their own but perhaps carry a suitcase from their parents' home to their husband's, to take flying lessons would have been an unthinkable expression of individuality. The decision to fly was frought with fears, not only of airplanes, but of herself. She had been taught to deny her capability. Whether or not she wanted to be a man, she usually did not wish to alienate men or the community in which values and attitudes were shaped by them. Even Amelia Earhart, while she was learning to fly in 1920, wore men's flying clothes not only because they were practical in the windy cockpit, but also because they made her less conspicuous at the flying field. She also purposely kept her hair long enough so that she would not seem "eccentric." And pilot Jacqueline Cochran made conventional femininity and its trappings her other profession.

But once a woman broke through the barriers of finance and fear, by involving herself in flight, she never could return to traditional perceptions of her place in the world. Such a young woman became one of the top stunt pilots in Pittsburgh during the mid-1930s. Her rise to fame illustrates the internal war fought by women who braved the air.

In 1934, while Earhart was preparing for her flight between Honolulu and California and Cochran was about to enter her first air race, nineteen-year-old Teresa James, who designed wedding sprays for her family's flower shop, was lured to the Wilkinsburg airport on the outskirts of town by a handsome young flier named Bill. Teresa was terrified of flying. When her parents had taken her for a balloon ride as a child, she had crouched in the bottom of the basket and not looked once. Yet every Sunday, Bill and his hearty group of fliers took off in their airplanes and flew to the mountains for a picnic. Every Sunday night, Bill's winsome passenger vowed she would never again put herself through another picnic trip. Then, suddenly, Bill left for Chicago to take a brief job as a flight instructor. Heartbroken, Teresa continued to haunt the airport.

One day, a flying buddy of Bill's suggested she surprise Bill when he returned by learning how to fly. Teresa braved five flying lessons and soloed. Instead of Bill, she began daydreaming about airplanes. When word came that Bill had got married in Chicago, Teresa did not even care. Her goals had changed. She entered the Buffalo Aeronautics Institute, as the only woman in her class, got her instructor's rating and learned to teach acrobatics and inverted flying tactics. On returning to Pittsburgh, she became a flight instructor at Johnson

Airport and earned $50 a weekend, a prodigious sum in the mid-1930s, as a barnstormer. The pretty twenty-one-year-old brunette became the spectacle of the air show with a specialty she developed. One Sunday afternoon, awestruck spectators counted twenty-eight spinning, plummeting turns before she pulled her plane level, only feet from the ground.

Six years later, after almost 3,000 hours of flying experience, Teresa James became one of the first women pilots to be recruited by the Army Air Forces.

If Amelia Earhart inspired women to seek independence and competence, Jacqueline Cochran represented success to many young people during the murky, dust-laden Depression years leading up to World War II. Like Cochran, they experienced an acute and desperate need for escape, and in flight sought the euphoria of freedom.

Growing up in the dry, southeastern Colorado prairie town of Granada, Alma Velut never went farther than fifty miles from home. Her father, who ran the local barbershop and billiard parlor, could not afford a car. In high school Alma took a secretarial course and on graduation in June 1940 left Granada for Pueblo, a hundred miles away. She found a job in the typing pool of a building company. One day an Electrolux salesman invited her to go out to the airport with him to meet a pal who was a flight instructor. He thought she might be impressed. She was, but not by him. Transfixed, she watched the small airplanes take off and land. From that day, she began to save quarters—and dream.

As Jacqueline Cochran was preparing for her transatlantic flight in the lend-lease Lockheed Hudson bomber, Alma Velut took the money she had saved and went to the airport for her first flying lesson. She could only afford fifteen minutes in the air at a time, she confessed. But she was determined to become a pilot. After each lesson, she wrote herself notes in her logbook, in which times and flight locations were recorded; she wrote them in shorthand so no one could read her thoughts.

Unlike Jacqueline Cochran, who in 1932 had soloed after three days of intensive instruction at Roosevelt Field, Alma Velut needed six months of quarter-hour lessons to solo. She barely had time to practice a landing. But every weekend, Alma sat in the hangar talking with pilots. Anyone who wanted company on a flight immediately got an excited volunteer. Every second in the air, as pilot or passenger, was flying experience for Alma. One man flew her all the way to Colorado Springs. She felt she was really going places.

By the spring of 1942, Alma had earned her license and was even able to afford a part-ownership share in a small Piper Cub. (The entire plane at that time cost $1,000.) She named it the *Challenger* and in it took on the whole state of Colorado. That two years later she would be flying five-passenger U.S. Army Norsemans across the United States and Canada would have been beyond her wildest dreams.

On a wheat farm twenty-five miles out of Wichita, Kansas, another young woman was dreaming of escape. The second of six children in a German immigrant family, Lucille Friesen spoke no English until she was twelve years old. A loner, the tall, dark-eyed child would not join her sisters playing dolls, but rejoiced when her father took her out to a tiny airfield where on Sundays barnstormers dropped in to delight the farm families. Every once in a while, her father let her go for an airplane ride. For days afterwards, she would be even dreamier than usual.

As Lucille approached graduation from high school, her older sister got married. But being a farmer's wife and having babies was not for Lucille. She had to get out, she said to herself, and *do* things.

Against her parents' wishes, Lucille moved to Wichita to attend business school. When a job appeared on the bulletin board for a stenographer in the bookkeeping department of Cessna Aircraft, she raced to the placement office. "I'd like that job," she said with uncharacteristic ebullience, "I just love airplanes." As she was one of the school's top students, she was sent to Cessna for an interview.

As Jacqueline Cochran was receiving her third Harmon Trophy in May 1940, Lucille Friesen took her first salary check straight to the airport and signed up for flying lessons. Soon, she had her license and owned part of a Taylorcraft Cub. On summer evenings when she and her friends would go flying after work, Lucille felt she never wanted to come down again.

Two and a half years later, and by then rated a commercial pilot, Lucille Friesen was recruited by Jacqueline Cochran to fly for the Army. When the telegram came, she related the news to her employer that she had to leave her job to report in two weeks for duty as a military pilot. Though Cessna made Army airplanes she would soon be flying, her boss would not give her serviceman's leave. Instead, before her eyes, he dropped a pink slip in her record folder prohibiting that she be rehired after her wartime assignment.

Lucille soon forgot her anger. She was up above the clouds for hours in the pilot's seat of a B–17.

There were also young women who refused to be left behind when the boys-next-door took to the skies and left home for greater adventure. When Margaret Kerr was a sophomore at East Central State College in her home town of Ada, Oklahoma, parties were getting dull. All the men gathered on one side of the room to "hangar fly," their hands looping and soaring through the air as they exchanged flying stories. A lot of the young men in Ada flew. It made sense. There was not much else but sky between Ada and anywhere else. When a government flying program ran for the spring semester in 1940, Margaret signed up. Brows in Ada raised, but her father, whose signature she needed for the application, was enthusiastic. Margaret had mumbled an inclination to become a lawyer like him, but he felt that the courtroom was no place for a woman. He did, however, buy her a small airplane and became her most insistent passenger. He did the navigating.

Soon after getting her pilot's license, Margaret was invited to a dance by an Ada boy who had left for New Mexico Military Institute, in Roswell. As NMMI was over 500 miles away, she doubted a weekend jaunt was worth the long train trip. "That's no problem," her father said. "Fly."

Margaret took off on a bright, clear Friday afternoon. Her 60-horsepower Cub coupe had no radio, but she followed the railroad tracks, which pilots called the "iron beam." After flying hundreds of feet above the panhandle of Texas for what seemed like hours, she began to see the land below her turning from beige to clay red. The sun was getting red, too, as she headed toward the distant Rocky Mountains. The field at NMMI was nothing but a cow pasture, her date had informed her. That did not worry Margaret. She flew out of lots of rutty cow pastures in Oklahoma. But there were no lights, and if darkness fell before she arrived, she would never find her destination. She began to worry that she had followed the wrong railroad tracks.

Just when it was so dark she could barely make out anything on the ground, she spotted far ahead a small group of tiny throbbing lights in the form of a rectangle. As she got closer, she discerned the dark outlines of campus buildings. She then realized that the four points of mellow light were bonfires. As she landed neatly amid them, young cadets in uniform came running out of the darkness to the plane. Their anxiety, and hers, forgotten with her triumphant arrival, they extinguished the fires and escorted the famous young woman who had flown herself all the way from Oklahoma to the dance.

If many flew to escape the geographical and experiencial confines of small-town reality, mastering flight also represented an escape from entrenchment in the past. At that time, flying was considered the most adventurous link to modernity and assured one's active participation in the decades to come.

When Kay Menges turned twenty-eight years old in the fall of 1940, she looked in the mirror and decided she was getting too old to wear a long bob. A new school year had just begun, she thought, always a time for reflection. She enjoyed teaching health and physical education at her progressive high school in Teaneck, New Jersey, owning a car and having her own apartment. But she was getting close to thirty and marriage was not imminent. "What will keep me in touch with the next generation?" she asked her mirror.

The next day, Kay looked at the list of adult education courses at her high school. Under "Aviation" were Aerodynamics and Engines. She signed up for both. A month later, she drove to Teterboro Airport for her first flying lesson.

Her flight instructor, however, was doomed to live in the past, especially when it came to teaching a woman student how to fly. He sat Kay in the back seat of a 60-horsepower Cub with dual controls, and he climbed into the front seat. As he shouted instructions back to her, Kay took off and flew once around the traffic pattern. As she was coming down for her first landing, he suddenly grabbed the stick, landed the airplane and taxied almost full-throttle to the side of the runway. Kay watched, bewildered, as he threw open the cockpit door and vomited into the marsh. Then he had her taxi back to the hangar. The lesson was over.

The next summer, she met an attractive man named Frank, who owned a communications company and an Aeronca airplane. Frank had soloed over Guantanamo Bay, Cuba, during World War I. During the course of their courtship, Frank learned she had taken three abortive flying lessons. "Why did you stop?" he asked. Kay mumbled something about the instructor's not exhibiting much faith in her flying ability. "That's ridiculous, a smart, athletic girl like you. I'll teach you how to fly."

Kay soon learned that flying came as naturally to her as tennis, basketball and field hockey. By the end of the summer, she had earned her pilot's license, and was on her way to a lifetime of flying and a lifelong partnership with the man who had faith in her.

Two years later, in the summer of 1943, Kay Menges would recall her first flight instructor. By then she was selected by Jacqueline

Cochran for a top-secret experiment to prove that women could be as cool-headed and competent as men when flying under fire.

As WAR BEGAN TO SPREAD throughout Europe, and while their elders held the line of neutrality, the sons and daughters of the Great War agitated for action. France, the Netherlands and Belgium had fallen to the Germans, Britain was under virtual aerial siege, and America's young yearned to take an active part against the growing nemesis of Nazism. This generation had watched the world tied closer together by airways and felt new affinities with the nations whose culture they shared. To them a threat to Europe was a threat to their own dreams of the future.

In Washington, the War Department, too, was looking to the skies. It was inevitable, after all the advances of the 1920s and 1930s, that airplanes would figure significantly, should the United States join the expanding world conflict. Thus, with war looming, organized programs were hastily instituted which would provide opportunities for women to learn to fly. Amelia Earhart would not have dreamed it possible, but the new programs would soon allow Jacqueline Cochran to create the largest women's air force in the world.

WHEN THE graduation rehearsal of the San Diego State College class of 1940 was interrupted for a few announcements, one made Jean Landis, a twenty-one-year-old physical education major, jump partway to her feet. A government-sponsored pilot training course was being offered to the first twenty people who signed up at the dean's office. "My chance!" Jean whispered to her classmates sitting around her in the middle of the gymnasium. "I've always wanted to fly."

Jean Landis had been born on her parents' homestead in El Cajon, then a dusty dot in the desert just east of San Diego. When the Landises had left Missouri and come to California at the turn of the century, they had lived their first five years in a tent house with wooden sides and canvas top. By the time Jean was born in 1918, her father, who worked with the local farm bureau, had built a redwood house. But there was no plumbing or electricity installed until Jean was of school age.

As a child, Jean watched airplanes fly over her house and thought, "Oh, what a thrill that must be, to be up there!" Now she was being offered the chance to find out. If her mother could leave her family's three-story mansion and maids back in Kansas City to conquer the California desert, Jean thought, she was enough her daughter to conquer the air.

The minute the rehearsal was finished, the tall blonde physical education major tore off her cap and gown and ran, jumping over sagebrush, down the campus canyon. She was the first student to reach the dean's office.

Flying lessons sponsored by the United States government were an almost incredible flash of luck to thousands of college students. For a $40 lab fee, most of which paid for a physical examination and flight insurance, they could earn a private pilot's license at government expense. In addition, college credit was given for the course. After much debate in the Civilian Aviation Authority, sponsor of the program, it was decided that one student in ten could be female.

The Civilian Pilot Training Program, as it was called, was the first time the government had paid much attention to aviation. In 1938, the armed services consisted of fewer than 5,000 military-trained pilots. Civilian pilots in America numbered only 21,000. Most of these had learned to fly at small, struggling aviation companies, usually made up of an incorrigibly enthusiastic pilot-owner, a flight instructor or two, and a couple of Cubs. The companies barely survived; Sunday airplane rides, a few charter trips and even fewer students were all most of them could boast of. The civilian aviation industry, primarily a network of shacks at the fringes of America's local airports, was still economically in the barnstormer era. As for the Army Air Corps, only seven years liberated from the Army Signal Corps, not much more could be said.

As the war came closer and the government projected how many pilots the nation would need, actual figures fell appallingly short. To build a reservoir without undue alarm, given prevailing isolationist sentiments, war planners in 1939 jumped on a heretofore wild idea posited by an owner of a small Utah aviation company who had just been appointed to the new Civilian Aviation Authority: Why not use existing teaching facilities—colleges and universities—for ground school instruction and contract with local aviation companies to provide flight instruction?

For fiscal year 1941, Congress authorized a $34 million budget for the Civilian Pilot Training Program, the equivalent of $7 billion today. By the end of the program, 400,000 Americans had become pilots at government expense. The CPT program was like a shot of a new wonder drug to the commercial aviation industry. It also helped win the war.

The one-in-ten ratio allowed to females was gratefully filled by women students across the country. Like their fellow men students, most of these women could never have afforded to learn to fly on their own. The alacrity with which Jean Landis signed up for the CPT program was equaled by Vivian Gilchrist, the little girl now grown in Jamestown, North Dakota, who could at last live the adventure books she had read in her tree. Another CPT student was Iris Cummings, the young Olympic swimmer turned aeronautics student at the University of Southern California; yet another was Margaret Kerr, whose flight from Ada, Oklahoma, to Roswell, New Mexico, would be the happy result of her CPT studies.

But with this unexpected wartime opportunity, others who might never have thought of flying found a lifelong enthusiasm as well as a skill that would soon be sought by their country. Elaine Harmon, a bacteriology major at the University of Maryland in 1940, felt starved for extracurricular activities. There was no tennis team for women, nor a golf team—the two sports she enjoyed and excelled in most.

Early in the fall of 1940, Elaine saw a notice in the *Diamondback*, the campus newspaper, that a CPT program was being formed at College Park Airport. At last, the university was offering a sport for which she did not have to sit in the bleachers.

To Margaret McNamara, a sociology major at the University of California at Berkeley, the CPT program was a chance to distinguish herself. She learned of it in a letter from her brother Robert, then teaching at Harvard Business School. He had been unable to pass the physical because of his poor eyesight (a myopia he retained, when, as Secretary of Defense under Presidents Kennedy and Johnson, he helped lead America into its thirteen-year Vietnam War), and he enclosed an application form.

Margaret could not resist the attraction of doing something he could not. Four years her senior, Robert had practically run the Berkeley campus singlehanded and Margaret had suffered from the kid-sister image long enough. When she went to the dean's office in the fall of 1940 to apply for the CPT program, however, she learned that the university already had its quota of female students—two. But she was now determined to fly. Among the four all-women's colleges across the country that were allowed to organize a CPT program was Mills, in nearby Oakland. (The others were Adelphi, in Garden City, Long Island; Lake Erie, in Painesville, Ohio; and Florida State, in Tallahassee.) Margaret mustered her courage and called the president of Mills. Her plea was so persuasive that, though not a student there, she was allowed to enroll in Mills' CPT program.

In Sioux Falls, South Dakota, that same year, Loes Monk became Augustana College's top flying student, on a dare. When Loes enrolled in the venerable eighty-year-old liberal arts institution, she joined the band and formed a singing group with two other women students to help pay her tuition. In 1939, no one on campus, most of whom came from Dakota wheat and grain farm families, had much money. So little, in fact, that the college's CPT program the following year almost did not get off the ground.

Her sophomore year, the popular blonde coed was dating a handsome football player named Bill. One of Bill's friends had signed up for the CPT program but, he told Bill and Loes despondently, one more student needed to apply in the next two days or the program would not go forward. Bill turned to Loes and suggested that she be that student. "They can't have any girls," Loes said with reasonable conviction. "Why don't you?"

"You wouldn't get me in an airplane if you paid me," said Bill, a strong statement in those days, "and they *can* have one girl." Bill and his friend dared Loes to sign up. Loes gave the ultimate excuse. She did not have $40. "Okay," said Bill, "if you show up at the administration building tomorrow morning, I'll show up with the $40." They shook hands.

Loes felt quite safe, walking over to the administration building the next day. Where was Bill, or anyone, going to come up with $40? (In 1940, $40 would buy 130 pounds of rib roast.) As she entered the building, Bill was leaning casually against the wall. Grinning, he handed her a fistful of ten dollar bills.

In shock, Loes Monk signed up for the CPT program. Within two months she was the most enthusiastic flying student in the college. She graduated first in her CPT class and the flying school at the Sioux Falls airport received so much publicity from their pretty student that her instructor offered his Stearman bi-plane to fly for free, any time she wanted to. Bill, a "flying widower," had long disappeared from her social calendar and she never did learn where he got the $40.

Bill did not know it, but he had contributed to the war effort a pilot who would train scores of navigators and bombardiers for combat. With graduation approaching in June 1943, Loes successfully auditioned her singing group before Lawrence Welk, who was recruiting talent for a road show to entertain America's military bases. Her family was outraged; they forbade her traveling around the country with an orchestra. But Loes refused to go back home to the farm. She sold her harmonium to Lawrence Welk, to pay for her train fare, and two weeks after graduation, was on her way to Sweetwater, Texas, where women pilots were training to fly for the Army Air Forces. Not until she had earned her AAF wings six months later did her parents forgive her for running away.

The lucky women in the CPT program did not always find a clear flight path. Adele Beyer was one of two women accepted for the twenty-student CPT class at Grinnell College in Iowa. Ground school instruction was held at the college during the day. But to fly, students had to rise at dawn and go to the airport to fit in a lesson before classes began on campus. Adele was in a bind. The women's quad was locked up tight until 7:00 A.M. Her impassioned pleas to the dean of women to let her out earlier for her flying lessons were to no avail. "We cannot jeopardize the safety of the other women students," the dean told Adele sternly.

The six-foot sophomore had graduated from high school holding several individual scoring records in basketball. She rode horseback, played clarinet, saxophone and trombone and was runner-up in the Iowa State singing championships. Her freshman year, Adele had organized women's basketball teams and founded an intercollegiate tournament at Grinnell. She was determined that the dean of women was not going to keep her from flying airplanes.

Every morning, Adele and her female flying buddy waited for the night watchman to make his rounds at 6:00. Then they jumped out the basement transom window of the women's dormitory and ran off to join the cars going out to the airport.

In spite of the dean of women, Adele Beyer would become commander of one of the largest female test-pilot squadrons in the country when the nation went to war.

At first, the CPT quota allowing one woman in ten was only valid for the CPT elementary course leading to a private pilot's license. Women were barred from highly selective advanced government-sponsored courses teaching acrobatics, cross-country navigation and flight instruction. Then in the fall of 1940, a spot landing competition was held in Seattle, Washington, which was open to all CPT elementary graduates. The prize to the winner was a scholarship to the advanced acrobatics CPT course. Much to the consternation of officials, the contest was won by a University of Washington home economics student named Barbara Erickson. After correspondence with Civil Aeronautics Authority officials in Washington, Barbara was allowed to take her prize, and advanced CPT courses were opened to women. One of the beneficiaries of her breakthrough was Jean Landis, who performed so well after her race to the dean's office that she was one of the five from her San Diego CPT class, and the only woman, to be chosen to continue government-sponsored advanced flight training to become a flight instructor.

By the summer of 1941, however, the War Department could no longer stand by and let civilian aviation authorities run their pilot training program. It was now clear that the war was soon going to be on in the United States and every pilot would be needed in the military. A pledge was required of all CPT students that they would serve in the armed forces. And as of July 1, women were out.

The Civil Aeronautics Administration (now a part of the Department of Commerce) squirmed under the strident criticism of the General Federation of Women's Clubs, the Women's National Aeronautical Association, hundreds of newspaper editorials and the ever vigilant Eleanor Roosevelt, who demanded an explanation. The CAA, with guns at its back, finally limped forward to face its critics with an official statement:

> It is generally recognized that male pilots have a wider and more varied potential usefulness to the armed forces than female pilots. We have had to make changes with this in mind, and during the past year our Program has, of necessity, been closely integrated with the needs of the Army and Navy aviation constantly in the foreground. To limit our training to young men is only another in a series of steps based on this philosophy.

Though the CPT program may have had a dual purpose at the beginning, to boost the aviation industry and train pilots for an emergency, by the summer of 1941 aviation in America was no longer "civilian." Since there was no military organization for women pilots, no more women would be trained to fly at government expense.

Like many young women on campuses across the country when the axe fell, Lydia Lindner, a senior at Brooklyn College, was only a few lesson hours away from passing her advanced acrobatics CPT course at Roosevelt Field, where fourteen years before Lindbergh had taken off for Paris. Most women CPT students across the country on July 1 packed their flight bags and left their airports for good, even though they were in the middle of their training. A few were able to complete their CPT courses, paying for them with borrowed money or by washing down and fueling airplanes at flight schools—airplanes their fellow men CPT students were still flying on the government.

Lydia Lindner, however, was lucky. She had so few hours left that a lengthy and outraged article appeared in the New York *Daily News*. Her Congressman and the CAA received repeated pleas from Brooklyn College and the CAA-contract flight school. Her fellow students also rallied to her cause. Not only was she an excellent and popular pilot, and the only woman in the course, Lydia's jalopy was their sole transportation from their homes back and forth to the airport. Lydia was allowed to finish her course. Two years later, she would prove her "potential usefulness" to the armed services in an artillery training flying mission that would almost kill her.

Though women were now barred from enrolling as students, those who had managed to get through the CPT instructor's course beforehand were avidly recruited by the CAA, which was desperate for instructors for the men who continued. The War Department did not object to women instructors. In fact, to assuage critics after women students were banished, both government groups sought to publicize the fine job women instructors were doing. When Barbara Erickson, the spot-landing champion in Seattle, became a flight instructor and graduated all six of her male CPT students as private pilots in 1941, General H. H. "Hap" Arnold, chief of the Army Air Corps, and CPT founder Robert H. Hinckley, now Assistant Secretary of Commerce for Air, were sent to Seattle to salute her before press cameras.

With the changeover to an all-male program, the CPT began to issue uniforms and pay students a monthly stipend, and, to reflect its new status, its name was changed to the War Training Service. The program also began to specialize. Students who could qualify physically were referred to Army and Navy air cadet programs to train as combat pilots. Those who could not, or were beyond the twenty-seven-year age limit, were given enlisted reserve status, thus plucking them from the draft pool. These pilots received further CPT/War Training Service instruction so they could serve in noncombat roles, such as liaison pilots with the artillery (like the original World War I scouts), service pilots with the Ferry Command, and as CPT/WTS instructors. Thousands of men went the WTS route in order to fly during the war. By far, however, the program's main responsibility was to supply flight instructors.

In 1942, CPT/WTS instructors were teaching all Army Air Forces primary, or first stage, cadets at AAF flight schools across the country. But by the end of 1943, the CPT/War Training Service had done its training job so well that the Army Air Forces, amply supplied with instructors and trained combat pilots, terminated the program completely. Like the women dropped in midtraining in 1941, thousands of men were now "released" from the CPT/WTS. But there was a difference. Unlike the women, the men lost their enlisted reserve status and were draftable. And in 1944, the government program that had given so many young women the opportunity to fly would take on the mission of grounding women for good.

The CPT program, limited as it was for women, had nevertheless made its contribution. Before it began, only 675 women held pilot's licenses. By July 1, 1941, when it ended for women, over 3,000 had become pilots. On special assignment to the War Department shortly thereafter, Jacqueline Cochran would discover this highly educated, enthusiastic group of flight-trained women and begin to formulate a plan.

When orders from the CPT program and the War Department began to pour into the Piper Aircraft Company in Lockhaven, Pennsylvania, another opportunity opened up for women to learn to fly. Its assembly line suddenly alive with Piper Cub trainers and liaison planes, Piper began to lose workers to more highly paid factory jobs in the big cities. Like many manufacturing company presidents, William "Papa" Piper started to recruit women. Though he did not pay as much as other defense plants, he had a persuasive selling point. He believed that his workers made airplanes better if they could fly them. Hiring instructors, he offered flying lessons to workers for $1.12 an hour, the price of the gasoline, and gave them an hour off the assembly line each day to fly.

After one ride in a small plane over Philadelphia, nineteen-year-old Anne Shields was hooked on flying. After many entreaties, her mother let her use some of her college money to begin lessons. As gasoline was rationed, every Saturday at nine o'clock in the morning, Anne boarded a trolley car in Philadelphia, took it to the end of the line, got on a bus, took it to the end of its route, and then walked two miles to the Souderton, Pennsylvania, airport. Though she did not return until seven at night, she was not always able to fly. Either the weather closed in, or a priority student from the CPT program took her flying lesson time slot. By the end of the summer of 1942, though she switched to another flying school thirty miles from her home in the opposite direction, Anne had obtained only eight hours of flying lessons. Then the East Coast was closed to civilian flying 150 miles in from the Atlantic coast as a wartime security measure. Anne was in despair.

A comic strip in the Sunday newspaper run by Camel cigarettes gave her new hope. It featured girls ferrying airplanes for Piper Aircraft upstate in Lockhaven. Anne wrote to Piper and got a response almost immediately from the personnel manager. Young women were indeed flocking to Lockhaven from all over the country, living in boarding houses, working in the factory and flying. If Anne could pay the train fare, she would be hired. All she needed to bring were her tools.

Anne did not know what a screwdriver looked like, but she did go to the men's department of Montgomery Ward and buy a pair of men's work pants and several blue workshirts. Soon she was sitting on her suitcase in the stairwell of an overflowing train bound for Lockhaven. Her mother, widowed when Anne was three, bid her a teary farewell. "If you get homesick, you just come right back," she told her daughter. Anne's brother, exempt from the draft as the sole surviving son, waved gleefully, hoping that Anne's foray from home would chart the way for his own. He was envious. All the young men in their Philadelphia neighborhood had left to join the service, and his kid sister was beating him to the opportunity.

Anne never did need any tools. Instead, she was assigned as the clearance officer at the Piper company airport. Every time a worker canceled a lesson, Anne stepped into the slot. After her halting beginning as a flying student, she was able to earn not only her private license but also her commercial pilot's license—which requires two hundred hours in the air—in only nine months.

By the spring of 1943, "Papa" Piper was again losing factory workers, this time women, to an Army Air Forces women pilots training program. But he did not forget them. Every Christmas during their active duty with the AAF, a carton of coveted Hershey bars would arrive addressed to them at their air bases, sent by "Papa" Piper himself.

IN NEW YORK CITY, an entire organization of women sprang up from the concrete, nurtured by the wartime spirit to serve—a society group founded solely to provide flying lessons.

On June 24, 1940, a twenty-five-year-old file supervisor at Standard Oil Company named Emily Chapin read her *Herald Tribune* on a commuter train from Larchmont, New York, into Manhattan. Every day the news from Europe convinced her further that America would soon be at war. Now the French had surrendered to Hitler.

For weeks, she and her friends at Standard Oil, mostly secretaries, had discussed what they would do for the war effort. They had worked hard to get where they were; they were financially independent, had pension plans and paid vacations, and they were proud to be working girls. But Emily sensed that there was now a chance to do much more with her life. Her younger brother Richard had joined the Merchant Marines and she, too, yearned to go overseas. But it was not so easy for a young woman. She had written to the American Friends of France, the Red Cross and the Military Training Camps Association. So far, possibilities seemed slim. She also began a Red Cross course in motor repair, but the need for chauffeurs and ambulance drivers had not materialized as in the last war. Passports to Europe, she was told, were harder to get, even for international organizations, as the turmoil increased. She also wrote to Jacqueline Cochran, who had been speaking out to women to take stock of their skills that might be of use to the war effort. Emily was determined to take advantage of something.

That morning in June, she saw an item in the paper that interested her. A group calling itself the Women Flyers of America was holding a rally at the Plaza Hotel that evening for young women who wanted to learn to fly airplanes for preparedness. Emily arrived at the office very excited and tried to convince her friends to come with her. They thought flying was going a little too far. Women were still guests of honor at society luncheons just for flying across the country as passengers. Emily resolved to attend the Plaza meeting alone.

The Plaza ballroom was aglow that evening, as the organizers of the Women Flyers of America—two professional pilots, a theater publicist and a New York model—sat on the dais and nervously watched young women enter the room. Successful women pilots organizations in the past, like the Ninety-Nines, were affiliations for those who had already earned licenses. The Women Flyers were different. Organizers had signed contractual guarantees with two flight schools that their membership would be large enough to merit discount flying lessons. Before the Plaza meeting, the Women Flyers of America, belying its grandiose name, numbered only about a dozen. How many career girls parading the streets of Manhattan wanted to become pilots was far from certain. After launching a flurry of publicity in the New York papers, organizers only hoped that their rally would not be a humiliating failure.

For weeks they had worked getting endorsements from prominent New York women; banker Jessie Q. Mott; Mrs. Floyd Bennett, wife of the famous Naval aviator; Katharine Hepburn; radio commentator Polly Shedlove Martin; choreographer Sara Mildred Strauss and Broadway actress Ada May. They also found Madame Ogilvie-Druce, the former Edith Berg, who had been the first woman to fly, with

Wilbur Wright as her pilot, in 1908. These illustrious names gilded the brochures which, with application forms, were piled on tables set up in the rear of the ballroom. Inside each brochure was the Women Flyers of America slogan: "Airmindedness for Sport, Profession and Emergency." As former New York City Mayor Jimmy Walker, who had greeted Lindbergh and Earhart when they returned from their pioneering transatlantic flights in 1927 and 1928, approached the dais to take his seat as a sponsor of the group, the organizers looked around the ballroom with amazement. When the doors finally closed for the rally to begin, over 500 women had come to join the Women Flyers of America.

As Emily listened to the speeches she realized her opportunity had finally arrived. "Aviation offers women the chance to play a new and great role in the defense of America!" The rhetoric infected the audience, who began to clap and cheer. "C'mon, gals, it's time we got ready to *do* something for this war that's surely coming," said Genia Novak, a professional model and membership vice president. "We think that flying is the answer. There are a lot of jobs we could do to release men to fight at the front—as mail pilots, pilots of passenger planes, air ambulance pilots. We could fly war planes fresh from the factories to military fields." She had a professionally modulated voice, in command of its rousing urgency. "Are we going to stay at home and knit mufflers like our mothers?"

Metropolitan Opera star Hilda Burke was presented an official Women Flyers of America flight cap. It was made of pure white leather and lined in sapphire blue.

As the rally dispersed, Emily Chapin joined the crowds of young women which gathered around the tables at the rear of the ballroom, paid her $5 membership fee and the first installment of her $250 flying costs. As Jacqueline Cochran had done eight years before, Emily took her vacation at an airport. And as full-scale Nazi air attacks began their relentless rain on Britain in mid-July 1940, Emily Chapin soloed in a Piper Cub over the potato fields of Long Island.

In October 1940, Emily Chapin became the second member of the Women Flyers of America to earn her pilot's license. She and her flying friends from Nassau Airport celebrated with beer and fried chicken at their local hangout, called The Spot. As "I'll Never Smile Again" was played for the tenth time on the jukebox, Emily's male flying buddies joked, "You've got your license, *now* what are you going to do?" "I'm ready for anything that comes along," Emily responded. A few months later, she bought a share in a 40-horsepower Cub she dubbed the *Mousetrap*. She flew on weekends and studied for her commercial and instructor licenses. She and some friends dreamed of having their own airport someday.

But fifteen months later, Emily Chapin would have the answer to her friends' question. It would come from Jacqueline Cochran.

VI

On the Precipice of War: An American Woman Joins the Battle for Britain

Emily Chapin stood in front of Rockefeller Center looking up at the bulging bronze statue of Atlas, kneeling, as he held the world above him. Her short brown hair whipped around her face in the winter wind. It was February 3, 1942, and as battles raged from Libya to Bataan, the world was far less stable than Atlas made it seem.

Riding up in the elevator, Emily was well aware that the imposing reputation of Jacqueline Cochran, whom she was about to meet, was interfering with her calm. After seeing the article in the *Herald Tribune,* Emily had written her immediately. Cochran was recruiting twenty-five American women pilots to fly for the Air Transport Auxiliary in Britain. The ATA needed pilots badly after over a year and a half of air combat. Fifty women, from Britain and several other Allied countries, were already ferrying warplanes from factories to RAF bases. For their eagerness and dependability, these women pilots had become known, affectionately, as "ATA-girls."

Emily had almost abandoned hope of going overseas when Cochran's offer appeared in the newspaper. She now had to convince the famous aviator that a year and a half of weekend and vacation flying in her 40-horsepower Cub, the *Mousetrap,* qualified Emily Chapin, a file supervisor at Standard Oil Company, to make a distinguished contribution to the Allied war effort in Britain. She did not know that civilian pilots like her were the backbone of the ATA.

As the elevator rose, Emily unbuttoned her coat and recalled that pivotal day, December 7, 1941. She had been listening to the New York Philharmonic concert on the radio as she put books and records in cardboard boxes. The next day, she was going to move in with Gertrude Holzer, her best friend from their days at Dana Hall School. After graduation, Emily had gone on to Wellesley; Gertie had married. Now, Gertie and her husband had just bought their first house, in Rye, New York, and Emily had offered to rent one room and contribute her furniture to help fill some of the others. When the concert was abruptly interrupted by the news that Pearl Harbor had been attacked, Emily stood stunned, a pile of books in her hands. One of her first thoughts was of her brother, Richard, who had finished officer's training school in the Merchant Marine and whose convoy was right in the middle of the Atlantic. But she had another concern: what war would do to her flying. As a woman, she assumed she would be grounded completely. The opportunity before her now, in Rockefeller Center, would have been the most preposterous of pipe dreams on December 7.

Finally the elevator door opened onto the thirty-fifth floor. As Emily walked into the offices of Jacqueline Cochran Cosmetics, Inc., a young woman at the reception desk greeted her in a southern drawl. Her name was Mary Nicholson. She was a pilot, too, she said, as she showed Emily through a large oak door into Cochran's office.

The door closed behind her. Cochran's office was decorated in eighteenth-century antiques. Lining the walls were lipsticks, pressed powders, bottles of lotion and perfume in blue packages with the identifying silver scrollwork of Cochran's "Wings to Beauty" cosmetics line. Compared to the striking blonde woman who rose from behind the desk and extended her hand, Emily felt very unimpressive. Since she used most of her $35-a-week salary on flying, she spent little on clothes or make-up.

"Tell me about your flying experience," Cochran said in a friendly tone after Emily had settled into the chair facing her. Nervously, Emily began to tell her interviewer about her 250 hours in the *Mousetrap*. Her mouth was so dry that every four or five words she had to swallow. Cochran, who understood "dry mouth," reached in her desk drawer and pulled out a roll of Lifesavers. Emily took the roll and continued her saga, sucking intermittently.

"I'll hold your name on the list," Cochran said finally. "When you get another fifty hours, you'll hear from me about a departure date. Glass of water?"

Emily shook her head. "What about the questionnaire I filled out last summer about women pilots being used by the U.S. Air Corps?" she asked.

"There isn't going to be any military flying for women in the States," Cochran said firmly.

On her way out, her ordeal over, Emily chatted with Mary Nicholson. Mary, too, planned to join Cochran in England as soon as the recruiting was finished. The two young women agreed to meet for lunch.

Over the spring, Emily clocked hours flying for the newly organized Civil Air Patrol, members of which were the only civilian pilots allowed in the air over war-restricted coastal zones. From Mary Nicholson, she learned that small groups of Cochran recruits were leaving for Montreal, to be screened by the ATA, and then sailing for England. She and Mary had been assigned to the last group. While Emily waited impatiently to be called by Cochran, however, the reality of war struck home.

On April 30, an article appeared on the front page of the New York *Sun*. It told of twenty-four Merchant Marines who survived eight days in a lifeboat on a ration of two biscuits and a cup of water a day. Their unescorted merchantman had been torpedoed only 400 miles from New York by a German U-boat. One of the ship's officers, last seen still aboard, had celebrated his twenty-eighth birthday the day before the attack. He was Richard Chapin, of Larchmont.

"Losing one child is enough," Emily's friends counseled over the following weeks. Ferrying airplanes in German-strafed Britain seemed like a terrible risk. Her mother, to her relief, did not agree. "This is all the more reason you should go," she told Emily.

On July 2, the telegram finally came from Canada: Emily was to report to Dorval Airport, Montreal, on Monday, July 6.

Gertie Holzer helped Emily put her two suitcases, hatbox, flight briefcase and Victrola in the car Sunday evening. At the train station, Emily gave a salute and they laughed. Suddenly the two friends fell sobbing into each other's arms.

As the train pulled out of Grand Central Station, Emily found herself wondering what a blackout was like, or a bomb shelter with dozens of strangers. How would she feel up in the same sky with a squadron of Messerschmitts? She cleared her mind with one thought: at last she was going overseas, to be *paid* for doing what she loved most, to fly.

MONTREAL'S Dorval Airport was dizzying with activity as British Overseas Airways Corporation, which ran ATA operations, prepared ATA pilots for service in Britain. Some of the pilots spoke other languages, but Emily noticed many American men who, like her, were eager to join the war effort overseas. Only a few United States military pilots had yet been sent. The few women on the field were Cochran's earlier recruits. Emily immediately met Opal Anderson, who was about to sail for Liverpool. Opal, who owned her own airport in Chicago, had 2,600 hours in the air as a commercial passenger and transport pilot. Emily, with only her *Mousetrap* adventures, wondered if she was in over her head. That afternoon, however, she began to collect her daily subsistence pay of $10, far more than she had ever earned before, and took a grip on her confidence. If she was being paid as a professional pilot, she would perform as one.

On the flightline, she was greeted by Captain Robert Smith, an ATA instructor to whom Cochran's last recruits had been assigned. His other women students had already arrived: Mary Nicholson; Myrtle Allen, a teacher from New Jersey; and Roberta Sandoz, a newspaperwoman from California. "Call me Bobby," said the tall, brown-haired reporter. Emily liked her ready, dimpled smile immediately. They all followed Captain Smith out on the ramp to their airplane—the 550-horsepower Harvard. It was similar to the U.S. Army Air Forces' advanced trainer, the AT-6 "Texan," he explained. Pilots called it the "Yellow Peril." If the women pilots could master the Yellow Peril, the ATA felt they could qualify on any of their single-engine airplanes, in particular the Hurricane and Spitfire fighters. Emily stared at it in awe. It made the *Mousetrap* look like a toy.

Captain Smith sat each of them in the cockpit for an hour as he went over the instrument gauges, gasoline mixture and throttle controls, the propeller switch handle, and several other knobs and dials. Emily feared she would never remember everything. "There is an easy formula for takeoff," he said. "Just think HTTMPPFG: Hydraulic landing gear, throttle tension, trim, mixture, pitch, petrol, flaps, gills—in this case, a carburetor." He pronounced the last word "car-byu-rator." Sounds like a cinch, Emily thought dubiously. The formula, if any, for the *Mousetrap* was "Fill it up and go."

On August 15, Jacqueline Cochran's last recruits had mastered the Yellow Peril, and were in St. John's, New Brunswick, aboard a Norwegian freighter that was bound for Liverpool. Now that she was about to sever ties with her home continent for the first time, Emily found herself a bit shaky. She felt no easier about the upcoming sea voyage. The freighter was armed with antiaircraft guns, but they would cross the Atlantic unescorted, their young captain told them. The ship went a fast fifteen knots and having a battleship escort would only slow it down. He had made twenty-eight crossings without incident from German submarines. He could outrun them.

Emily did not know how fast her brother Richard's merchantman had been, but the havoc wreaked by undetectable German U-boats even on escorted convoys along vital Atlantic shipping routes had been devastating—1,000 ships lost since January. She hoped the U-boats would not notice, or at least would ignore, one small freighter carrying a cargo of tinned food and four women pilots going to join the British ATA.

Fog accompanied the freighter protectively for eleven days. On August 26, when Emily and Bobby Sandoz climbed up on deck after breakfast they noticed there were more crew members than usual standing watch. The antiaircraft guns were manned. From the front and rear masts, two large kites flew 1,000 feet into the sky. The crew told them kites would prevent planes from diving at the ship. Then, seeing their two passengers' anxiety, they pointed off to starboard. Not ten miles away was Northern Ireland. By noon, they could see Scotland and the Isle of Man. Emily and Bobby watched the shoreline lovingly all day.

As the sun was turning red, the crew, with great ceremony, dropped anchor. Then a tug came alongside. Up the ladder climbed a greeting squad of men and women Air Transport Auxiliary pilots. Within an hour, the new recruits were packed and chugging toward the Liverpool docks. Emily looked back at the freighter on which she had crossed the Nazi-patrolled Atlantic. It looked like a pebble among the craggy gray battleships strewn across the harbor.

Once through customs they sent cables home. To pay, they held out their hands and the operator took bills and gave them change. In their preoccupation with flying, it had never occurred to them that they should learn British money.

The ATA pilots took them to a canvas-top truck that looked like a covered wagon. Along each side were wooden boards for seats. Jostling through Liverpool, her feet on her luggage, Emily looked out the open back of the truck. A block of buildings would suddenly stop, make way for jagged parts of walls and piles of cement and dirt, and then resume with solid, unharmed façades. The people walking along the streets looked strained and mirthless.

The truck stopped in front of a huge hotel. Still vibrating from the ship, Emily and Bobby were shown to a suite of rooms, with sofas, chairs and a bathtub big enough for an entire family. At dinner, the peas were fresh, but the ice cream tasted of paste. Emily realized she was in a country at war. There was no sugar or cream to make ice cream.

AS HER last four recruits stepped off the train in London on Wednesday evening, September 2, 1942, Jacqueline Cochran stood on the platform with an entourage of American Army officers. In a flurry of military activity, luggage was piled into an Army truck and the young women placed in waiting cars. As they drove through the city, Emily looked out the window, trying to catch her first glimpse of London. A blackout, she discovered, was like being in a cave. Only the vaguest outlines of buildings were visible against a pitch-black sky. Finally the cars stopped and the recruits followed the drivers into an elegant foyer. The man at the reception desk looked confused. "Oh, you want the Park Lane Hotel," he said. "It's next door." The blackout was so total that many before them also had mistaken the two entrances.

The recruits were taken upstairs to their rooms. When London miraculously appeared with the morning light, they found themselves gazing out their windows onto Piccadilly and Green Park. All that day, Jacqueline Cochran's cars drove them to various offices to be photographed, registered and given ration books. Finally, they were measured for uniforms at Austin Reed, Ltd., a swank London clothier. In the evening, Cochran invited them to her house for cocktails and to a London club for dinner. She sat Emily and Bobby on either side of her. To make conversation, Emily said the meal was grand. Cochran smiled mischievously. One of her colonels brought in extra supplies, she told her guests. "When one of my pilots got appendicitis, I got her the best medical attention in England," she went on expansively. "Nothing's too good for my girls." Emily returned to the Park Lane a bit heady with wine and warmed with a feeling of well-being.

On Sunday morning, she and Bobby finally had time for a walk by themselves along Hyde Park. Suddenly, a siren began to wail. The sound that had regularly pierced the days and nights of London sent chills through their flesh. People around them seemed to change direction instantaneously. Emily and Bobby moved with the crowd through the doors of a large building and down a set of stairs. The two Americans looked at each other, their eyes wide with anticipation. They were actually in a bomb shelter with the British; they waited, expecting it to quake with German bombfalls. Instead, Emily was soon aware of a rank odor of sweat, stale tobacco and urine. Ten or fifteen minutes passed before another siren sounded and people moved toward the stairs. As the outside air washed over her, Emily was almost disappointed not to see any smoke. Cochran, who had driven them around the more battle-scarred sections of London, had assured them the bombing now was sporadic. After ferrying a bomber there in the summer of 1941, she told them, she had watched the air attacks from her blacked-out hotel suite at the Savoy. One missed too much in shelters. Emily decided that the next time a siren sounded in London, she probably would be safe staying above ground.

That night, most of the twenty American ATA-girls, already through ATA orientation training and out flying, came into London from their bases across England to attend a dinner for Jacqueline Cochran at Grosvenor House. As the new recruits met their predecessors, they looked longingly at the handsome blue ATA uniforms and officers' stripes. The women who had arrived early in the year seemed practically to be combat veterans. Before Germany turned its attention eastward toward Russia, bombardment of England had been relentless. The British had so few aircraft that when an air alert sounded at an aerodrome, ATA pilots had to scramble into all airplanes on the field and take off—unarmed. Sometimes, they delivered them to safety at an aerodrome not on alert. Often, however, they simply circled, helplessly, until the Luftwaffe had spent its last bombs. In 1941, the ATA had lost one of Britain's most famous women pilots, Amy Johnson Mollison. It was presumed that a German fighter peeled off and shot her down over the English Channel. Even if they salvaged their airplanes from destruction on the ground, landing them on bomb-cratered airfields "undamaged" tested every skill they had as pilots. Emily listened to their stories with awe.

Cochran's seasoned recruits were flying two- and four-engine bombers or the famous British fighters, the Hurricane and Spitfire. Emily was thrilled to think that she, too, would soon be doing the same.

After they had all been seated, Cochran rose and, in the barest hint of a Georgia accent, began to talk. Now that her last girls had arrived safely, her job in England was finished. She was leaving next week, to attend to business that needed her attention back home. She knew, Cochran went on, that they would continue to do a fine job. She wanted everyone to keep in touch with her.

After dinner, they repaired to Cochran's house for drinks served by a waitress. The new recruits floated back to the Park Lane, with two questions in mind—why had General Sherman, a century earlier, said war was hell, and why was Jacqueline Cochran leaving Britain so soon?

THEIR GALA welcome was, before long, a glittering and incongruous memory. Emily, Bobby, Myrtle Allen and Mary Nicholson boarded a train to Luton, forty-five minutes to the north, where, after a three-month technical course and cross-country navigational training, they would become Third Officers in the ATA. ATA pilots, having no barracks, were often taken in by English families in return for their ration books and a weekly stipend. The recruits were split up, and Emily was billeted with a London solicitor named Foster, his wife and five-year-old daughter. Mrs. Foster agreed to give her breakfast and dinner for two pounds a week. As Emily could get lunch at the aerodrome canteen for a shilling, she presumed her seven pounds a week cadet salary would be plenty. Mrs. Foster showed her upstairs to her room, which overlooked a meticulous garden behind the house. Emily could have a bath every night if she liked, Mrs. Foster said. When they called her down from unpacking for dinner, Mr. Foster had set out the cocktail tray. Emily knew she would be happy there.

At the Luton aerodrome the next day, Emily rejoined the others to see the 120-horsepower open-cockpit RAF trainer, the Miles Magister, or "Maggie." They were issued gas masks, boots, helmets, goggles and wool-lined flying suits that had zippers everywhere and huge pockets on each thigh. When they tried on their new gear, Emily and Bobby remarked that they looked like Martians. In their first technical lecture they were given a letter from the ATA Commander, G. d'Erlanger. Two items were underlined: the *secrecy* with which they must treat their every movement, and the importance of delivering every airplane *undamaged.*

Saturday morning, her fifth day in Luton, Emily was sitting in a meteorology lecture. The airplanes flown by the ATA, even the fighters, had no radios or navigation aids other than a compass. Such equipment was saved for planes going on combat missions. Emily listened intently. She had never seen such rapid changes in weather as she had noticed in England. She would wake up to sunshine, then fog would cover the field. Gusts of wind would carry the fog away, then it would rain. By afternoon tea, the sky would be clear again. She wondered how anyone in the ATA ever completed a flight without navigational instruments.

After two weeks of technical lectures and map study, they finally got to take up the Maggie. Emily's instructor, Mr. Peabody, who wore four ribbons on his jacket, had been flying for twenty-six years. At last, for the first time since checking out in the Yellow Peril a month before, she was back in the air. England from 500 feet looked like all the photographs she had seen—green, gently rolling fields, and hedgerows enclosing cozy clumps of sheep and cows. She found that her many trips over New England made navigation here easy. Then one afternoon, she flew her Maggie right into an English rainstorm. The ground and sky merged into a single gray mass. Mr. Peabody let her fumble several minutes with her maps in the narrow cockpit and make a few turns, her compass lolling drunkenly in front of her disbelieving eyes. Then the veteran of a quarter-century in the air took over the stick. "What will you bet we hit home right on the nose?" he asked. The next material object Emily saw was the aerodrome runway five seconds before they touched down on it. As they walked to the pilot's hut, she stared at her instructor in awe. "I think I'm strictly a fair weather pilot," she said, her voice wavering. "We all are," he said.

By October, Emily received her first mail from home. Sentences, often paragraphs, had been cut by the censors. Among her first letters was an enthusiastic note from Marion Mackey, a friend from New York with whom she had learned to fly thanks to the Women Flyers of America. Enclosed was a newsclipping. Secretary of War, Henry L. Stimson, had announced the formation of a women pilots group in the U.S. Air Transport Command, under the direction of Nancy Harkness Love, called the Women's Auxiliary Ferrying Squadron, or WAFS. A women pilots' training program was also being formed and Marion had applied. The training program was to be directed by Jacqueline Cochran. Remembering Cochran's assurances that no women would be flying in the United States, Emily could not help feeling somewhat betrayed. On arriving in England, she had signed an eighteen-month contract with the ATA.

But Emily felt British already. She had learned that "a jolly good show" would be the response from her instructor after a successful solo cross-country flight, and not to ask for a "rest room" at an aerodrome because it would have only a bed in it. If her instructor took "a very poor view" of a cockpit maneuver, he meant, "What the devil did you do *that* for?" "Circuits and bumps" was not a risqué dance, but takeoff and landing practice in the traffic pattern. Calling them "stooge" told how the ATA felt about copilots—they could scarcely afford to have them, except on a training flight. With the help of a "torch," Emily was learning to make her way through the blackout by herself, though batteries for it were hard to find. And she now knew to protect her dilated pupils when she arrived at the door of a pub. "*They're* not in England," she said, folding up Marion's letter.

Only one thing about her experience was troubling—the perpetual chill. Her evening bath promised by Mrs. Foster could only be three-

and-one-half inches deep, due to wartime fuel restrictions. As enjoyable as the Fosters were, she found that she became hoarse chatting with them too long in the unheated parlor after dinner. Excusing herself politely, she would retire to the coke-burning stove in her room. One evening she stood so close to it that she burned three large holes in her bathrobe. By early November, her hands and feet were so British they had chilblains. She cut an article about the weather from a London paper that read:

> The only thing that can be said for it is that it has made us what we are. The nation that can stand the British weather is not very likely to be daunted by Hitler. As for the Americans, they will probably get as hardened to it as anybody can; and when they go into action it may be the people at the other end of their guns who suffer from cold feet.

Emily certainly hoped so.

In early October, as General Bernard Law Montgomery prepared his Eighth Army to attack Rommel's forces at El Alamein, Egypt, to check their advance toward the crucial Suez Canal, Emily Chapin took her first cross-country check ride testing her navigational knowledge of southern England. Miraculously encountering no rainstorms in the area, she felt she had done well. "Okay, you'll do. In fact," the ATA check pilot said as she opened herself to a compliment, "you come up to our standards." Taken aback by his understatement, she found she liked the British expression for passing a check ride even less. "You have passed out," he said finally, in an almost positive tone. She would now train on the 650-horsepower Hart. If she "passed out" in it, she could at last wear the blue uniform of the ATA. However, she would be allowed only four hours of instruction in the Hart. Ferry pilots were supposed to be versatile.

The jump from the 120-horsepower Maggie to the Hart meant many new gadgets to master. Emily discovered with glee that one of them was a retractable heater. Now that it was mid-October, even her Martian-like winter flying suit was becoming inadequate for protection.

On Sunday, October 11, Emily passed out in the Hart, after three hours and forty minutes of dual instruction. That same afternoon, she was sent up in a 200-horsepower liaison plane, a Tiger Moth, for an hour of circuits and bumps. Never before had she flown an airplane without first receiving instruction in it. She landed after an hour feeling very versatile indeed; she could wear her uniform with confidence.

The next day, Emily and Bobby, who had passed her flight test in the Hart the day before Emily, walked out of Austin Reed, Ltd., into the brisk October sunlight in dark blue caps, skirts, and reefer jackets with a "USA" patch on each shoulder. American soldiers hooted at them along the street.

For the last stage of training before receiving their wings and ranks as Third Officer, the ATA-girls had to complete twenty-five solo cross-country trips. During one, weather forced Emily to land on an RAF base. At luncheon, she was the only woman among 200 RAF pilots. Beneath the flirtatious banter she was distinctly aware of the strain on the faces of her fellow fliers, most of whom had logged over two years of defensive air operations, first in France, and then over their own country. The faces of even the youngest pilots, who could not have been over twenty-one, were lined. As Emily walked back toward the aerodrome the fog cleared and she saw small groups of pilots lounging on the grass, resting in the brief rays of the afternoon sun. Only the "Mae West" life preservers strapped tightly around their bodies betrayed the reality that at any minute they would be called to rush for the waiting line of fighter planes poised for takeoff across one end of the airfield.

For their first overnight leave in London, Emily and Bobby headed straight for the Park Lane, where they had first been put up. After paying the bill, which took a week's salary, they had more appreciation than ever for Jacqueline Cochran. They decided in the future to find other facilities. One day they walked past a townhouse on Charles Street with an American flag in front. They entered to find a Red Cross club for nurses, run by Mrs. Anthony J. Drexel Biddle, Jr., wife of the United States Allied Ambassador. The cafeteria served good coffee, Coca-Cola, and peanut butter. In the corner was a jukebox that played for a penny. The club also had free movies and for four shillings they could stay in dormitory rooms upstairs. Army musicians played in the evening for concerts and dancing. The American ATA-girls were welcome, they were told. They now had a home base in London, and, they soon found, a guaranteed date with any of the United States officers who flocked there for dinner.

Behind the blackout curtains, London theaters and concert halls were bright, and Emily learned she could hop a train into London on days the weather washed out any flying. Wartime performances began at 6:00, so she could be back in Luton by midnight.

One rainy afternoon, she and Bobby decided to go into London to a show. But soon after they settled into a compartment filled with U.S. Air Corps engineers they changed their plans. By the time they reached London, they all had become great pals.

After an afternoon of sightseeing and shopping for wool undershirts, Emily and Bobby met two of the engineers at a Piccadilly bar for drinks. Both men were from the South, as most of the American men Emily was meeting seemed to be. After several rum and Cokes, they went searching in the blackout for a place to dance. They found Prince's, which had a swing band led by a "Canadian," who spoke in what Emily recognized as a Bronx accent. When they closed Prince's well after two, the last train back to Luton had left. The engineers gallantly invited the ATA-girls to stay with them in their hotel. When the lights were turned out, the women pilots fell into one single bed and the two engineers fell into the other. At 6:00 A.M., Emily and Bobby dressed quietly and left to catch the 7:05 train to Luton. They had to stand all the way back, feeling miserable. Fortunately, fog swirled around Luton aerodrome all morning, leaving the ATA-girls shakily, if gratefully, on the ground.

The next day, fully recuperated, they took up their first Fairchild 24. The high-winged Fairchild was used as an ATA taxi plane to transport pilots to or from a delivery of a bigger combat airplane. It had a 175-horsepower engine, room for four and, much to Emily's delight, an enclosed cabin. Though she enjoyed her cross-country navigational practice flights, Emily was impatient with training and wanted to be put to work. When the news arrived that the Germans had moved in to take over unoccupied France on November 11, Emily felt a new surge of commitment. For the first time since arriving in England eight weeks before, as an ATA taxi pilot, she could at last be of service to the Allies.

IT WAS already Thanksgiving. Emily went to London to attend a service at Westminster Abbey. The famous church was chillingly cold inside. Maneuvering through the American gold braid and popping flash bulbs, she found a seat near the sanctuary. As United States Ambassador to Britain John P. Winant gave a Thanksgiving proclamation, Emily was awed by the ancient statues and pillars around her. The sermon was delivered by a man who sounded like the announcer at a boxing match, but halfway through the final hymn, "America the Beautiful," Emily shivered, not only from the cold but from a sudden gust of chauvinism.

When she returned to Luton on the 7:05 the next morning, Emily was told to take her Fairchild instructor up in an airplane she had never flown before, a 220-horsepower Avro Tutor. After twenty minutes of stalls and simple flight maneuvers, he asked her if she was happy. Surprised, she said, yes and no. She loved the flying but she hated the cold. The commanding officer wanted to see her, her instructor told her with a smile. Though she had only completed fifteen of the required twenty-five cross-country trips, she was going to get her wings. The ATA needed fighter pilots, and all cadets whom instructors thought capable were being accelerated into operation as Third Officers. Her uniform would now bear its first stripe.

Their preliminary training completed, the American ATA-girls were dispersed from Luton to ferry pools all over England to be "seconded," or given temporary duty, to learn ATA operations as taxi pilots. Bobby and Emily were sent to the 15th Ferry Pool in Hamble on the southern coast of England. After Christmas, they would be sent to ATA headquarters in White Waltham, east of London, for training on the Hurricane and Spitfire, and their next promotion to Second Officer.

On December 3, Emily and Bobby checked into an inn, just off the Hamble River, called The Bugle. The bar had a real log fire, and like everything else in England, Emily found it very quaint. The heaters in their rooms worked by dropping pennies into a slot. In the morning, they ordered bacon and eggs for breakfast. On the rations of an English family, Emily had eaten six eggs since leaving America.

The 15th was an all-woman ATA ferry and taxi pool, headed by Captain Margot Gore. Captain Gore was second in command to Pauline Gower, founder of the women's division of the ATA, and was one of few women qualified to fly four-engine bombers. Miss Gore's assistant, Margaret Smith, Emily learned with amazement, had been a dancer before the war. The 15th's main responsibility was to ferry new Spitfires from the Vickers Supermachine factory in nearby Eastleigh. On Emily's first day of duty, she was sent in a Fairchild to follow two ATA-girls as they delivered Spitfires from Eastleigh to an RAF station near Dover. At last she was doing an actual job, even if just to bring home two pilots who had delivered airplanes to someone who needed them.

Flying along the southern coast around the port of Southampton, Emily was more aware than she had been in Luton that just across the Channel was the enemy. German bombing raids were launched on England from northern France and the Low Countries, only a matter of miles away. A thousand feet over aerodromes and factories all over southern England floated huge dirigiblelike balloons on cables, to discourage strafing. Though the scene looked like Herald Square at the end of Macy's Thanksgiving Day Parade, the bloated canvas puffs were deceptively devastating to the propellers of a diving airplane. To land, Emily had to circle the field until ground crews identified her airplane by sight (she had no radio to call into the field), and cranked down the balloons near the runway. Then she dropped in quickly before they were cranked back up again.

One day she was assigned to pick up a pilot in Plymouth. Enjoying one of the clearest days of the winter, she headed southwest out of Hamble and climbed up to 1,000 feet, gazing over the Isle of Wight and the Channel resort town of Bournemouth. Suddenly, she spotted two large planes speeding by at tree-top level below her. With a jolt of alarm, she recognized from their wing configurations and markings that they were German Messerschmitts. As she watched in horror for dark clouds of smoke to billow up from the city, she wondered, having no radio, what she would do if they turned back in her direction. "I've got no gun," she thought absurdly, having no idea how to use one. Suddenly, the Messerschmitts banked south over the Channel and disappeared. As her eyes darted over the coast, she could see no evidence of a hit. They were probably taking reconnaissance photographs, she realized, but her heart pounded all the way to Plymouth.

APPROACHING HER first Christmas away from her family, Emily had difficulty getting in the spirit. The undecorated shops in Hamble were empty of gifts. Good news from Africa that General Montgomery and Lieutenant General Dwight D. Eisenhower had driven the Axis armies back from Egypt, safely away from the Suez Canal, and into Tunisia, elicited her first hopes that the Allies might actually be winning. She also felt her first pangs of homesickness. Emily decided to report to duty Christmas Day. After she picked up a stranded Fairchild and brought it back to Hamble, the weather closed in. With Grace Stevenson, a Cochran recruit from Holderville, Oklahoma, whom they called the "Oklahoma Kid," and three English ATA-girls, Emily ate a Christmas dinner of roast rabbit at the "drome." That evening, they all ate another dinner at the cottage of one of the English women, this time of pheasant. The three Britishers argued that the bird had been hung too high for three weeks, but Emily found it delicious. After dinner, following the lead of her English colleagues, Emily marked Christmas 1942 with her first cigar.

Compared to their brief secondment at the small women's ferry pool at Hamble, ATA headquarters at White Waltham was like Piccadilly Circus. As ATA's central training pool, White Waltham, in Maidenhead, was constantly receiving and discharging pilots converted from one class to another. It struck Emily immediately as too hierarchical for her. (Every time she had flown in to refuel over the past few months she had bounced her landing.) The adjutant, a colonel, was a beribboned veteran of two wars whom Bobby Sandoz called "Laughing Boy," for his haughty, stern dealings with his pilots.

When Emily and Bobby arrived at their new billet, it looked, from the outside, like a grand old English house. Inside, it was like a barn. Their room had no heat, not even small heaters. Emily's hot-water bottle was arriving later by Anson transport with her luggage from Hamble, so she put on all her long underwear and socks and went right into bed to keep warm. Though it was three weeks after the winter solstice, Emily's first night in Maidenhead would seem like the longest of the year. Her bed was barely wide enough to turn over in, and when she did, the blankets fell off. She got up and put on her gloves and laid her overcoat on top of her. Then she felt the cold coming up from underneath, so she got up and put two of the blankets on the mattress. When she pulled the remaining blankets over her, she realized they were all only three and four feet long, and soon she was wrapped in a jumble of them. "We're better off than the Russians," she muttered, but she never took much comfort in comparisons. None too soon it was dawn.

When she and Bobby got up, they saw snow on the ground. The bus that was to pick them up and take them the two miles to the aerodrome had got stuck, so they had to walk. Emily's chilblains throbbed all the way. Arriving in the classroom, Emily waved to Mary Nicholson, who had arrived from her secondment in the north of England. But she could barely concentrate on their first technical lecture, which began with a discussion of superchargers.

At lunch, Emily and Bobby went straight to the billeting officer. "Oh, we're all cold," he said. "Is that your only complaint?" They spent that night in a hotel. Their room had a fire which kept them warm, at least one side at a time. The extra guinea a week, they decided, was worth paying.

The next day, rested and alert, they discovered that their technical instructor was very good and that the airplanes they were about to fly were among the most exciting in the world—the Hawker Hurricane and the Vickers Supermarine Spitfire, the fighters that had fought at Dunkirk, as well as against wave after wave of Nazi Junkers and screaming Stuka dive bombers, to win the Battle of Britain. Of the 3,000 RAF fighter pilots, Churchill had said that never before had "so much been owed to so few." For two weeks, Emily copied off the blackboard intricate diagrams of manual and automatic boost controls, different kinds of constant speed and fixed pitch airscrews and six types of carburetors. She filled a notebook with lecture notes. Class II, or Second Officer, technical ground school would culminate in two and a half days of written and oral examinations.

The oral examination on engines Emily dreaded most. The underlying rivalry between mechanics and pilots, she had heard, often surfaced when the pilot was a woman. When the day came, Emily sat down in front of her instructor, her teeth gritted for the verbal onslaught of, "How can you fly an airplane when you're so dumb you don't even know that!" Instead her exam went smoothly; as many of her questions were answered as his. Fighter deliveries were too vital to the ATA for each pilot not to be taken seriously, male or female.

After passing her last exam solidly, Emily celebrated her second officer status at one of Maidenhead's grandest restaurants with an ATA male pilot. Much to her surprise, the dour ATA headquarters adjutant appeared at their table and invited them for a drink. Unsure of how to entertain the upright colonel, Emily told him about the *Mousetrap*, which she had sold before leaving for Canada. She had just received a clipping picturing the tiny Cub in the parking lot of the Rye, New York, high school. Now famous because its former owner was flying in Britain, it was being used for aeronautics classes. For the first time, she saw the colonel smile. Then he asked her to dance. An accomplished athlete with several tennis trophies to her credit, Emily felt that how a man danced was an important criterion. When they sat down again, she decided the adjutant was not as staid and hierarchical as she had thought.

THOUGH SCHEDULED to begin flight training in the Hurricane after exams, Emily learned that, instead, she, Bobby and Mary Nicholson were to be seconded for ten days to a village eight miles from Leicester, called Ratcliffe, to deliver a group of Tiger Moths to be used in training RAF liaison and reconnaissance pilots. Eager to start flying fighters, Emily was disappointed. Moreover, she wanted to go anywhere but north, and she cringed to think where they would be billeted. A canvas-covered transport met their train in Leicester and drove through the darkening countryside. Then it turned into a tree-lined driveway and stopped in front of a palatial house. As the girls dragged their flight kits and bags out of the transport, the door opened to reveal a black-and-white-clad butler.

They were ushered, eyes popping, into a large wood-paneled room lined with portraits and perfumed by a log fire at one end. "Welcome to Ratcliffe Hall," said a handsome gray-haired man. He introduced himself as Sir Lindsey Everard, Air Commodore, RAF. Emily thought she was watching a movie.

The butler, whose name was Smart, served dinner at a long table lit with lamps decorated with pictures of horses. There were several decanters of wine on the table, but Emily was just as impressed by having a napkin, the first she had seen since the Park Lane.

After dinner, they were shown to their quarters atop the carriage house, a cavernous room covering more automobiles than Emily could count in a discreet glance. Each had her own room, with a washbasin and hot-water tap, and a radiator. Before they retired, Smart offered to clean their boots.

After her Tiger Moth delivery the next day, Emily returned to Ratcliffe Hall to join Mary Nicholson and the other ATA billetors for dinner, which was served at 7:30. While they were waiting, Bobby came in from her ferrying trip flushed with excitement. She had flown into an RAF base that was being reviewed by the King and Queen. Bobby joined the end of a line of pilots standing at attention. When the Queen had reached her, she had looked surprised and then smiled at her. Bobby had been the only woman in the line.

When weather prevented flying, Emily found that leisure was taken just as seriously at Ratcliffe Hall. The two tennis courts were out of doors, but an ATA pilot billeted there initiated her into the indoor game of squash, giving her the first exercise she had had since leaving her tennis-swim club in Westchester County, New York, the summer before. A Victrola and noteworthy collection of records entertained them in the evenings with interludes on the player piano. When Emily could not resist taking a roll of photographs to send home, one of the ATA men developed them for her in Sir Lindsey's darkroom.

But grand as it seemed to the American ATA-girls during their brief stay, Ratcliffe Hall had seen changes since the war. Sir Lindsey, a Member of Parliament, had long been a supporter of aviation in England. Ratcliffe, where they flew every day, had been his private aerodrome before he gave it over to the ATA for the duration, along with his home. At meals, Lady Everard hopped up to help Smart clear the table. Their daughter, Lady Newtown Butler, who was Emily's age, had gone to work as a driver for the ATA.

The evening of January 29, Ratcliffe Hall was bright behind the blackout curtains, as the Everards and their extended ATA family celebrated the news that Roosevelt and Churchill had met face-to-face in Casablanca in faraway Africa, and vowed to exact "unconditional surrender" from their Axis enemies.

DURING THE FIRST two weeks in March 1943, every day presented the most beautiful flying weather since the early fall. One by one, the ATA-girls were sent up for their first rides in the Hawker Hurricane. Emily's flight instructor, however, seemed not to know she existed. She had finished her navigation training in the twin-engine Oxford, and day after day she would walk past the Hurricane to take up a Harvard, the Yellow Peril she had flown in Montreal.

On March 11, Mrs. Biddle opened a house just for American ATA-girls across from the nurses' club on Charles Street in London. Emily went to the opening gala, attended by women's ATA commandant Pauline Gower and other ATA notables, and danced until the early morning. At 6:30, she got up to catch a train out to Maidenhead to report to the aerodrome. This morning, as she was throwing down cups of tea to wake up, her flight instructor finally noticed her. They went up in a Harvard for over two hours of cross-country heading changes, different kinds of runway landings and simulated engine failures. Fortunately, she was now so comfortable in the ship that she could almost fly it in her sleep.

She had just put her feet up in the pilots' room to eat a sandwich when he walked over to her again. Pointing out the window, he said, "That Hurricane is for you."

This was not the day she wanted to face 1,050 horses alone, she thought, as she walked up to the waiting fighter. In a single-seater, even the first takeoff was solo. She did her cockpit check very carefully, started the engine and rolled out onto the runway. As she opened the throttle, and as the plane accelerated, she was pushed back against the seat. After takeoff, to hold the proper rate-of-climb speed she had studied in ground school, she found that the nose of the fighter pointed almost straight up. The countryside seemed to race by under her.

Approaching the downwind leg of the circuit, she went through the landing check list like a sped-up movie; flaps down, mixture rich, prop pitch set. Then she put her finger on the undercarriage lever. It would not go down. She circled the field again trying to lock the lever into the down position. "I'll run out of gas before I'll belly-land my first Hurricane," she said to herself. For thirty minutes, she pushed on the lever. Suddenly it went down and she heard the wheels lock under her. She landed, handed the ship over to an engineer and walked, shaking, into the pilots' room. "Drinks on me tonight," she announced faintly.

The next week she was given her second stripe and did her first job as Second Officer Chapin, Class II ATA pilot. She delivered an 820-horsepower Miles Master to an ATA base. Compared to the Hurricane, the advanced fighter trainer seemed so tame that she made one of her best landings in months. When she walked in to have her delivery chit for the Master signed, a man by the desk was applauding. He was Mr. Peabody, her Maggie instructor from Luton, the veteran pilot who brought her safely back in the rain. "You did me proud," he said.

A letter awaited her back in Maidenhead from Marion Mackey, her Women Flyers of America friend, who was now at Jacqueline Cochran's training school for women pilots in Houston, Texas. Marion was flying twin-engine AT–10s and AT–17s, advanced Army trainers, and receiving hours of instrument training in an on-the-ground flight simulator called a Link trainer. To fly the sparsely equipped ATA airplanes, instrument training would have been superfluous, though after many close calls with marginal English weather conditions, Emily yearned for such help. " 'Taint fair," she grumbled, but Mr. Peabody's words still warmed her.

That weekend, Prime Minister Churchill told the British people they would know at least another year of war.

ONE SATURDAY afternoon during the first weekend in May, the Charles Street Red Cross publicity worker asked Emily and Bobby if they would mind being interviewed on "Red Cross Reporter," a radio show broadcast to the States. On their next day off, they were sitting at a large table in the BBC with a microphone in the center, across from commentator Lindsay MacHarrie. Emily put on headphones and read over the script. It was a bit glamorized, she thought. The earphones crackled and Emily heard a voice she recognized. He was a WCBS announcer, giving the morning news direct from New York. She suddenly felt she was going to cry. Then another voice broke through and said, "Calling London, are you ready?" The New York announcer introduced "The Red Cross Reporter," and MacHarrie began to talk.

Finally, he turned to his guests. "In this age, feminism is no longer an issue," he said into the microphone. "Women can vote; they can do almost any job, and most of them let it go at that, still cherishing their position as the gentle sex when there's a man around. But not the women of the British Air Transport Auxiliary." He asked Emily how she felt doing a man's job. "The ATA is unique because men and women are on exactly the same basis," said Second Officer Chapin. "You find that out when you're walking back from your plane carrying a parachute and kit, and there's often a completely unburdened male walking unconcernedly beside you." Bobby talked about the wonderful Red Cross club milkshakes, and in minutes, MacHarrie signed off.

As people in the studio continued talking back and forth to New York over the channel, Emily wondered longingly if her friend Gertie Holzer in Rye had heard her. She had. The day before, Gertie received a telegram from WCBS in New York. They were pleased to inform her, it said, that at 9:15 the next morning, "your son will be heard on the Red Cross Reporter Program."

ON MAY 6, Emily returned to White Waltham from ferrying a Miles Master to Wales. Next to her name on the training pool scoreboard in the dispatching office was the letter "S." She was at last to fly her first Spitfire.

The Spitfire had 200 more horsepower than the Hurricane. With one flight in the sleeker and more streamlined Spit, with its tapered wings, Emily joined many other pilots in believing it was the most beautiful airplane in the world. It even starred in a film called (in America) *Spitfire*, with Leslie Howard and David Niven, and Emily went to see the film just to see how she looked flying it.

Instead of the chunky, mullioned canopy over the Hurricane's cockpit, the Spitfire had a clear bubble. While the Hurricane could fly at 335 miles per hour, the Spitfire Mark IX could reach over 400, though the ATA flew them more slowly to break in their high-performance engines. Her first day, Emily got to take up two Spitfires and fell breathlessly in love.

For two nights, the 15th Ferry Pool gave its new Spitfire pilots a royal welcome at The Bugle, which rang with toasts to their return as Second Officers and to the news of victory in Africa, where the Western Desert Air Force was using the Spitfires the ATA-girls were delivering to force Rommel's Afrika Korps out of Tunisia and across the Mediterranean. Permanently situated at last, Emily and Bobby rented a cottage in Hamble, which soon became an ATA-girl microcosm of the Allied world. Among their roommates and visitors were Diana Burnato, daughter of a South African diamond mine owner; Maureen Dunlop, an Englishwoman who had grown up on an Argentinian ranch; Mardi Gething, the Australian wife of a British RAF pilot; and Vega Pilsudska, daughter of the Polish war hero Pilsudsky, who had flown herself to freedom in 1939 in a Polish Air Force plane. Representing South America, Chile's first woman commercial pilot, Margo Dualde, who had come to England after flying for the Free French, was always showing up with a bottle of port, an embassy attaché or a passing test pilot.

Emily's overseas experience was beginning to match her expectations for foreign contact. She was proud to be part of the largest Allied flying force open to females (outside of Russia). They all were flying as near to the edge of the war as a woman could get. When she landed her first Spitfire on an RAF base on the east coast, Emily saw an entire wing of thirty airplanes revved up for takeoff. She hopped out of the Spit and watched as they left the runway two-by-two and circled above the field until they were all airborne. Then, in formation, they headed east on business.

The Fosters had given her a history of England for Christmas and, every day, after picking up a new Spitfire from the Eastleigh factory, Emily flew at 200 feet over the spots she was reading about: Winchester Cathedral, the rocks of Amesbury, Stonehenge and the curved row houses of the ancient Roman spa of Bath. And every day she delivered Spitfires to RAF pilots, maintenance units and modification centers where instruments would be added, markings changed for camouflage, or flags painted on, designating the Allied country of the pilot—some would bear the Stars and Stripes. Sometimes fuel tanks were added for long-distance escorts of four-engine bombers bound for Germany. Every time she got into the cockpit, she thought, "This plane may be over Nazi territory tomorrow." The idea struck her with both a sense of pride and a chill of horror.

On American bases throughout Britain, which were growing in number as the Eighth Air Force was building strength, Emily discerned an air of excitement. The Eighth had been bombing Germany sporadically since the end of January 1943, and pilots now talked with hushed enthusiasm about a large-scale invasion of Europe sometime next year. Several of her countrymen would fly her Spitfires on RAF raids, to become the seasoned squadron leaders of the Eighth Air Force on far-off D-Day. Meanwhile, German industrial targets would be struck, and, daringly, by daylight.

Throughout the summer, Emily began to meet P–51 and P–47 pilots, freshly trained in the States, arriving in England in greater numbers by ship and bringing their airplanes with them. She argued heatedly about the virtues of British fighter aircraft. Usually, they all would agree that their airplanes were evenly matched. They were, after all, on the same side in the war.

"This is really more like it," Emily wrote home. "Lots of work, tired at night, but a feeling of a little accomplishment." She had been away a year. In that year, she had gone from flying a 40-horsepower Cub to being qualified in fourteen different airplanes, two of which were among the fastest in the world. She was also having an experience so broadening that she felt herself changing, like a ten-year-old who can almost feel her bones creak in growth spurts.

When she received a newsclipping picturing Marion Mackey standing at attention with her fellow graduates from Jacqueline Cochran's USAAF women pilots' training school in Houston, Texas, Emily's only envy was their suntans. Though it was June, her woolen winter uniform was still a month away from being packed in the closet.

One day, when Emily returned to her cottage after delivering two Spitfires, Bobby Sandoz met her with a look of ecstasy on her face. She was engaged to an ATA ground officer. Emily was amazed. Bobby had known him for only a month. Emily agreed to be maid of honor.

A week before the wedding, Emily ferried a Fairchild, an Oxford and, finally, late in the day, a Spitfire. England had gone on double daylight saving time and she was now flying exhausting ten-hour days. She took off in the Spit and as she left the aerodrome area, her hood blew open. Though she had not yet picked up the full 200-miles-per-hour cruising speed, the wind hit her in the face like a hurricane-driven steel wall. Crouching down in the cockpit, she turned steeply back toward the aerodrome. Tears blurred her vision as she took peeks out the side, but as she approached the runway to land, she realized she was at a slight angle to the field. Suddenly the Spitfire dropped hard on one gear, the wing raucously scraping the concrete. Then, like a goony bird, the plane bounced over onto the other gear, to scrape the other wing. Mortified and angry, Emily spent the entire evening filling out reports, among them a pilot's complaint about the defective hood latch.

Back at Hamble Friday morning, Captain Margot Gore told Emily that the accident committee at White Waltham wanted her to report for a flight check—her account of the Spitfire accident had been too "light-hearted." Emily hoped that Bobby's wedding, the Saturday before her check ride, would bolster her morale.

Friday afternoon, Emily went into London to spend the evening with the bride. They talked late into the night. Bobby contemplated the expatriation that her marriage would entail; Emily confided that she could see no further ahead than her next Spitfire or Hurricane delivery. Her future seemed a blur of excitement and confusion. On Saturday afternoon, August 7, Bobby Sandoz was married, as Emily stood beside her. After the reception, the happy couple went off for a week's "compassionate leave" in Oxford, and Emily went back to the Red Cross club to rest up for her check ride.

On Monday morning, she met the chief flight instructor, a British ATA captain, on the flightline at White Waltham. He made her fly for two hours, doing emergency flight maneuvers, complicated navigational problems, and every conceivable type of routine and emergency landing. "The accident committee was probably in a bad mood that day," he reassured her when they landed. "You were fine." He did not think the committee realized that Americans talk differently, even on paper.

Back in Hamble, savoring a rare evening of solitude in the cottage, Emily put on a new Debussy record and thought about her roommate. She wondered about a marriage based on a relationship only a few weeks old. And she thought about all the majors and captains she herself had written home about. Her most steady friendship was with a thirty-eight-year-old British ATA pilot with a great sense of humor. He was entertaining, world-traveled, and, best of all, interested only in her company, and no more. Early each evening that they spent together, he pulled out the most recent photos of his wife and five children. Among the American officers she met at the Red Cross club, many turned out to be married. But so many had "not been getting along well lately" with their wives, that Emily began to think that her parents, who had braved an unfashionable divorce in the mid-thirties and lived happily apart ever after, could teach the men of her generation a thing or two. Then she had gotten wise.

With the unpredictability of ATA flying, however, dates were almost impossible for her to keep. The men had the same problem. So she had learned to recognize those at the Red Cross club who, like herself, wanted spur-of-the-moment company at the theater and a good dance partner, a friend with whom to share the interludes of war. She was hardly bereft of company, she realized, but no one seemed to stick around very long. Nor did she, and she had never enjoyed her life more than in the last twelve unpredictable months.

IN EARLY SEPTEMBER 1943, Emily received a mimeographed letter from Jacqueline Cochran, written August 17, to all her ATA recruits in England. For five weeks, Cochran had been in Washington, she wrote, "on a very important job." She had been named director of all women pilots within the U.S. Army Air Forces. Her new women pilots group was called the Women's Airforce Service Pilots (WASPs), and her office was in the War Department's new Pentagon—"with lots of marble floors to walk." Five hundred WASPs were in training; twenty-five were graduated and out ferrying airplanes for the Army. Already, "her girls" were starting to fly twin-engine B-34 bombers, A-20 Havoc attack bombers and single-engine pursuits. "Naturally it is on an *experimental* basis," Cochran continued. "It is impossible to be on any job in Washington and not have a lot of politics develop." But if Emily or any of the others decided to come home, there would be a place for them in the WASP. "But I feel that I should warn you," she added, "it is just a matter of weeks before it will be militarized, which will mean if you come home, you will have to join the army." In closing she promised to send over a package of woolen socks.

Emily sat down and wrote her that as soon as her ATA contract was completed, she very much would like to join the WASP, at last to fly for her own country.

On September 8, 1943, the war took a decisive turn. That day Emily and an Australian ATA friend, Mardi Gething, arranged to ferry Spitfires to an RAF base near London because they had matinée tickets to *Love for Love*, with John Geilgud. After the play Emily met a date and all three went pubbing at a hangout for Australian pilots. Emily invited several back to the Red Cross club for a dinner dance. Eight of them piled into the same cab, and in the darkened streets of London, no one could tell who was who. One of the Australians on the top of the heap whispered sweet nothings into Emily's ear. The hand he was holding, however, was that of Emily's date, who relayed in her other ear every move and squeeze. When they all entered the brightly lit Red Cross club dining room, Emily was still in the dark as to who had been so keen on her. But the cab ride was soon forgotten. The dinner crowd that night was loud with shouts and toasts. The news had just been announced on the radio that Italy had surrendered to the Allies.

The autumn weather began to turn sporadic. Emily's hot-water bottle had been out of the closet and into her bed for a month. In between Spitfire or Hurricane deliveries, which often took only fifteen minutes, she sometimes spent two hours on the ground drinking tea with other ATA pilots, waiting for a howling gale to pass over the aerodrome. Unlike her home flyways of New England, in Britain there was usually an aerodrome underneath her if the cloud ceiling fell while she was in the air, though her navigational skills were taxed to know which one of the hundreds it was. As the days turned grayer in October, her average cruising altitude went down to about 500 feet. Though she carried a thirty-six pound parachute pack on every trip, she rarely got high enough for a parachute to do her any good. Flying so low at 220 miles an hour in a fighter, she had to know her topography well. England has 700-foot mountains.

One day Emily returned to Hamble after making only two Spitfire deliveries all day. Both trips, she had flown between fog to one side and rainstorms to the other. She had not enjoyed a comfortable moment on either trip. When she opened the door of the pilot's room, the other ATA girls met her, their faces solemn. Mary Nicholson was dead, they told her. That morning, she had flown into a mountain.

JUST BEFORE Thanksgiving, Irving Berlin's, *This is the Army,* began its tour of U.S. bases in Europe with a week-long engagement at the Palladium in London. Emily and the American ATA girl whom they called the Oklahoma Kid, Grace Stevenson, were able to get tickets. On its tour, one third of the house for each performance was reserved free for Allied troops. From *This is the Army, Mr. Jones,* to *This Time Will be the Last Time,* soldiers performed the entire show, including the orchestra, the Ladies of the Chorus and the hostesses of the Stage Door Canteen. The audience laughed uproariously. Emily, however, was surprised to feel lumps in her throat as the show went on, recognizing typical American gestures and expressions. Before the finale, Berlin himself came out on stage and sang his World War I classic, *Oh, How I Hate to Get up in the Morning,* with the audience catcalling and cheering. At the end of the show, the applause lasted many minutes, but when Berlin came out for a bow, everyone rose from his or her seat and burst into another Berlin classic, "God Bless America." Berlin began to direct the orchestra. By ". . . my home, sweet home," the entire audience of the Palladium—Americans, British, Dutch, Poles, Czechs, Frenchmen and everyone else—was weeping.

Grace and Emily boarded a milk train and sat on their parachutes in the aisle all the way down to Southampton. When Emily finally got back to her Hamble cottage, it was 2:00 A.M. She turned on her bedroom light to find her bed covered with gaily wrapped packages. In the exhilaration of the evening, she had forgotten that it was her twenty-ninth birthday. Her birthday messages were written in three different languages.

The day she flew her one hundredth Spitfire, Emily learned that her ATA contract had come up for renewal. In spite of Captain Margot Gore's flattering pleas for her to stay, after eighteen months in Britain, Emily realized she was eager to fly for her own country. Four of Cochran's ATA recruits had left to join the WASP. She was ready to go home. When she turned in her issued ATA gear, the matériel officer let her keep her overcoat for the voyage across the wintery Atlantic, a far safer ocean than it had been a year and a half before.

On February 23, the *Mauritania,* packed with soldiers and military personnel, sailed into New York harbor. As a Red Cross volunteer drove Emily up Sixth Avenue, she looked out the car window at the New Yorkers and compared them to the faces she had seen out the back of the canvas-covered transport in Liverpool. "There's no war on for these people," she thought in amazement. The Red Cross volunteer let her off on the corner of 50th Street by a Walgreen's. She dragged her luggage into a telephone booth, called Gertie Holtzer to come and meet her, and headed for the soda counter, where she ordered a hot fudge sundae, piled with peanuts. That evening, she and Gertie walked arm in arm toward Times Square, bought tickets for *One Touch of Venus,* with Mary Martin, and stayed the night in a hotel. It had central heating. Emily was home.

VII

Takeoff:
How Two Powerful Women Won Out Over a General

On a steamy summer day in early July 1941, Jacqueline Cochran, armed with President Franklin D. Roosevelt's letter of introduction, arrived at the War Department for her appointment with Assistant Secretary of War for Air, Robert M. Lovett. Upon her return to America on July 1, from ferrying a lend-lease Lockheed Hudson bomber to Britain, Cochran explained to Lovett, she had discussed with the President at Hyde Park a proposal for an organization of women pilots within the U.S. Army Air Corps. She did not know, nor did anyone, how many experienced women pilots were available across the United States, but she was willing to bring her own paid staff from New York to Washington to review the files of the Civil Aeronautics Administration. One of the initial flying duties she envisioned for women would be ferrying trainer-type airplanes for the recently established Air Corps Ferry Command, as women in Britain were doing so competently.

Jacqueline Cochran left Lovett's office as a "tactical consultant," and though she would not be paid, she and her staff would have access to office space in the Ferry Command section. The following Monday, she reported for duty. Chief of the Army Air Corps, General H. H. "Hap" Arnold, who in April had suggested she fly the Lockheed Hudson to Britain, welcomed her to Washington and introduced her to Colonel Robert Olds, the Ferry Command's commanding officer, with whom she would work for the next three weeks.

Cochran's staff of seven researchers began laboriously checking through over 300,000 files at the CAA. Finally, they located the records of 2,733 licensed women pilots, 150 of whom had over 200 hours of flying experience. Many qualified to be airline pilots, though women were not being hired by the major airlines. (In 1934, the first woman was hired as an airline copilot—Cochran's friend and future ATA flying colleague in England, Helen Richey, engaged by Capital Airlines. Within months, however, Richey was prohibited from flying most of Capital's major routes due to alleged "passenger complaints," and with her forced resignation, women would not sit in the cockpits of airliners for another thirty years.) But from the CAA records, Cochran did not know what types of airplanes or the extent of cross-country experience any of America's women pilots had flown. There was only one way to find out. She sent them all a questionnaire, with the urgent request that it be kept confidential, which asked how many hours they had flown, what planes, where, and whether they would be willing to fly for their country in the event of an emergency—in a war.

Within two weeks, she had received responses from 130 of the 150 commercially rated pilots who stated that they were not only willing but enthusiastic about the prospect of flying for the Army Air Corps. But from a sampling of replies from women pilots with fewer than 200 hours in the air, Cochran learned that hundreds were just as eager to join a women pilots organization. "I'll be available any time, anywhere," wrote Brooklyn College student Lydia Lindner, who, like many women college students, had just been removed from the Civilian Pilot Training Program flying rolls as of July 1. The overwhelming response from her questionnaire convinced Cochran that America had a large and previously unknown reservoir of women pilots who, with further training, could serve with the AAC to release men pilots for combat flying duty.

Meanwhile, from Colonel Olds, head of the Ferry Command, she obtained projections of how many trainer aircraft were on order for the AAC until the end of 1942. She matched them with her women pilots statistics and drew up a proposal.

On July 30, 1941, Cochran's proposal for an "Organization of a Woman Pilot's Division of the Air Corps Ferrying Command," was submitted to General Arnold. Since women were successfully flying in England and the Soviet Union, Cochran wrote, an experimental group of experienced women pilots in the United States might begin immediately flying small Army Air Corps trainer aircraft from factories to air bases. If, after a ninety-day trial period, this group proved women could do the job, they might form the nucleus of an AAC women pilots division. At this point, she wrote, over 2,000 licensed women pilots could be called upon, with very little transition training, to join such an organization. The number required, which Arnold could determine, should be commissioned in the Air Corps Specialist Reserve, and she should be retained as Chief of the Women Pilots Division. As AAC pilot personnel needs expanded in time of war, her organization of women would be ready to take on any domestic flying assignments to release men pilots for active duty.

At the time Cochran submitted her proposal several branches of the armed services were considering women's auxiliaries—the Army's WAAC was authorized by Congress nine months later and the Navy's WAVE in less than a year. These women's groups, however, were designed primarily to assist in service operations, a great many in clerical positions, so as not to "waste a good man" on a typewriter. These groups as proposed, and later as organized, stayed within contemporary society's norms, that is, women to act as "service personnel" for men. But the idea of a women's auxiliary of pilots was not yet acceptable to high command. Flying jobs were considered too important to give to women.

Upon reading Cochran's proposal, General Arnold disagreed that the pilot shortage would ever be quite so dire as to have to use an entire division of women. "Frankly," he later admitted, "I didn't know in 1941 whether a slip of a young girl could fight the controls of a B–17 in the heavy weather they would naturally encounter in operational flying." A month after he received Cochran's proposal, Arnold responded for his files: "The use of women pilots serves no military purpose in a country which had adequate manpower at this time." He wrote that there were so many men pilots in America, over 90,000 of them, and so few women, that he saw no reason that men could not handle both delivering the nation's airplanes and waging a war in them. In fact, in July 1941, the AAC had more pilots than airplanes. Furthermore, Arnold added, housing and feeding women pilots presented a "difficult situation" on air bases.

Before he dictated this memo, however, General Arnold called Jacqueline Cochran into his office. The time was not right for women pilots in the AAC, he told her in a friendly manner. He suggested, instead, that she accept the British Air Transport Auxiliary's request to recruit American women pilots for duty overseas. Britain's need was much greater. Thus, while she could perform a valuable and patriotic service to the Allies, she also might use such a group as an experiment to see if something similar might some day work in this country. Satisfied that nothing more could be done, Cochran left Washington and returned to her cosmetics company in New

York. Her arduously researched proposal would lie dormant for over a year.

JACQUELINE COCHRAN's was not the first proposal that General Arnold had turned down. More than a year earlier, in May 1940, while the Germans were marching toward Paris, a letter arrived at the Army Air Corps Plans Division which was at that time still organizing the Ferry Command. The Command was to be responsible for delivering new airplanes ordered by the AAC to training bases, modification centers, coastal ports of embarkation for shipment overseas, and, if necessary, flying military airplanes to combat zones around the world. The letter, addressed to then Lieutenant Colonel Robert Olds, who was directing the Ferry Command planning efforts, was from Boston pilot Nancy Harkness Love, who offered to recruit qualified women pilots for the new Ferry Command. "I've been able to find forty-nine I can rate as excellent material," she wrote, evidently in response to a previous inquiry from Olds. "I really think this list is up to handling pretty complicated stuff. Most of them have in the neighborhood of a thousand hours or more—mostly more, and have flown a great many types of ships. . . ."

Nancy Love's credentials to propose a women pilots group, though less sensational than Jacqueline Cochran's, were nonetheless substantial enough to get a hearing within the AAC. The daughter of a wealthy Philadelphia physician, Love was educated at Milton Academy, a Massachusetts boarding school, and at Vassar. Though only twenty-six years old, she had already been flying for a decade. Since 1936, she and her husband, Robert H. Love, had built a thriving aviation company in Boston called Inter-city Airlines, for which she was a pilot. But she had also flown for the Bureau of Air Commerce in a Works Progress Administration project to airmark water towers and rooftops with the names of towns and cities to aid pilots flying across the country. Love had also safety tested airplane innovations for the Bureau, among them the tricycle landing gear (now standard on most aircraft).

Olds read Nancy Love's letter with interest. He knew that the Ferry Command would be in need of good pilots, especially in the event of war when AAC combat forces started plucking away every man qualified to fight in the air. In this event, the proposed Ferry Command would have to fulfill its mission with pilots who were not in war-essential jobs (airline pilots or manufacturer's test pilots), who were over the twenty-seven-year age limit for combat, who had some minor physical disqualification for combat—or who were female. If Nancy Love could provide fifty experienced women pilots, Olds would be delighted.

Almost immediately, Colonel Olds made the suggestion to the newly appointed AAC Chief, General Arnold, that hiring women pilots be considered for the new Ferry Command. General Arnold was opposed to the idea then, too, just as he was initially to Cochran's later. Women might better serve in an emergency as copilots on domestic airlines, he told Olds, to release men pilots for service in the AAC.

As they made their pre-Pearl Harbor proposals to the AAC, in 1940 and 1941, both Nancy Love and Jacqueline Cochran assumed that the AAC was getting many individual inquiries from women pilots across the country. They were right. Typical was that of a young wife in Queens, New York, who wrote her local Air Corps recruiter. "Isn't there anything a girl of 23 years can do in the event our country goes to war, except to sit home and sew and become grey worrying? I learned to fly an airplane from a former World War ace, and I've forgotten how many parachute jumps I've made . . . if I were only a man there would be a place for me." Her cry was so poignant it made the New York *Herald Tribune*.

Though General Arnold had rejected the idea of using women in June 1940, the AAC continued to refer all such letters to Colonel Olds, now the head of the new Ferry Command. He was so beleaguered that in exasperation he recommended an AAC expression of policy, and at least a publicity campaign to encourage women to apply instead to Fiorello LaGuardia's new Civil Defense Agency Civil Air Patrol.

At that time, the Civil Air Patrol was recruiting women—not to fly, but rather to perform typical auxiliary functions, to organize and staff its operations offices. If a female volunteer happened to be a pilot, a memorandum from the commander of the 2nd CAP Region in New York stated, she would thus prove her ability "to work, to organize and to take orders." Then, if she could find her own staff replacement, she would be allowed to fly. Undiscouraged, over 2,000 women were flying for the CAP by mid-1942. In other words, a high percentage of the women with pilots' licenses acted as Jacqueline Cochran predicted—they served their country flying however they could. But they were prohibited from what was considered the CAP's most important wartime function—coast patrol. Authorities would not take responsibility for a woman pilot flying over the U-boats immediately offshore.

Then came Pearl Harbor. America had declared war, and it would be a war in which air power would be crucial. Immediate evidence of this shift in the War Department's thinking came when the title Army Air Corps was changed to Army Air Forces (AAF). Within the AAF, numerical projections, both of pilots and of airplanes necessary to fight a war in the air, multiplied many-fold. And as predicted by Ferry Command head Robert Olds, now a Brigadier General, the AAF combat forces began to take every available and qualified pilot out of reach of the Ferry Command. At the same time, the AAF was ordering an unprecedented number of new aircraft, which the Ferry Command would have to deliver. As early as January 1942, General Olds resuscitated the moribund proposals for using women pilots for ferrying duties, and advised Jacqueline Cochran that he planned to hire women pilots "almost immediately" on the same basis as male civilian pilots.

The timing for Cochran could not have been worse. After over two years of trying to persuade her government that women pilots could be organized in an emergency, the emergency was at hand and the Ferry Command was actually going to hire them. But rejected by the AAF, Cochran had now taken on other responsibilities. She had promised Lord Beaverbrook and the British Air Transport Auxiliary to recruit twenty-five women pilots, with three hundred hours of flying experience. Olds' pronouncement that he was hiring women to fly in the United States filled her with horror. Though her recruitment efforts had been endorsed by President Roosevelt and the Commanding General of the AAF, she was not having an easy time getting women with that amount of experience who were also willing to sign eighteen-month overseas contracts. If the American Army Ferry Command were plunging another bucket in the same precious pool, she would not be able to fulfill her responsibility to the British. She well knew from her research of the previous summer that only seventy-five to a hundred women had the kind of flying experience to ferry airplanes for the ATA or at home.

Cochran sat down and wrote a furious note to General Arnold, stating this argument. But her objection to General Olds' intention went beyond the potential for embarrassment in front of America's ally should she renege. Women pilots must not be integrated into the Ferry Command, Cochran reiterated to General Arnold, but should be a separate corps commanded not by men, but by a woman. Hiring women on the same basis as men, individually and without any organization, she warned, "will bring disrepute on the services before very long and injure the interest of the women themselves.

. . . His plan should be put on ice for at least six months, or my program for England should be stopped." Cochran knew she was in a policy dispute with Olds and that she had to be the more persuasive to General Arnold. "In addition," she continued, "it would wash me out of the supervision of women flyers here rather than the contrary, as we contemplated."

General Arnold sent Cochran's entire letter to General Olds. And, in a note of his own, as Olds' commanding officer, Arnold directed his Ferry Command chief: "You will make no plans or re-open negotiations for hiring women pilots until Miss Jacqueline Cochran has completed her present agreement with the British authorities and has returned to the United States."

That was that, at least for the moment. In late spring of 1942, Cochran went to England to oversee the integration of her twenty-five American recruits into the ATA. She anticipated that by mid-August, the last of the group would be screened in Canada, placed on a ship for Britain, and safely sworn into ATA service. Then, fresh from her British success, she would approach General Arnold again to get a group started in the United States.

But while Cochran was in England, Nancy Love found herself again in a position to offer to recruit women ferry pilots. In the spring, her husband, Robert, was called for military duty in Washington, as the deputy chief of staff of the new and rapidly expanding American Ferry Command, now called the ferrying division of the Air Transport Command (ATC). After the Loves moved to Washington from Boston, Nancy Love got a civilian administrative job with an ATC ferrying division operations office in nearby Baltimore. She commuted to her job by plane.

Along with its administrative reorganization, the ferrying division had a new cast of characters. General Olds was now head of the Second Air Force, and succeeding him was Brigadier General Harold L. George. Under George, as head of the ferrying division's domestic wing, was Colonel William H. Tunner. Together they faced the same shortages Colonel Olds had faced and worse. One morning in June, Colonel Tunner and Major Robert Love met at the water cooler at ferrying division headquarters. Major Love was voicing his concern about whether his wife had arrived safely at her office. When he explained about his wife's aerial commute, Colonel Tunner had an inspiration. "Good Lord," he exclaimed to the young officer over their water cups, "I'm combing the woods for pilots, and here's one right under my nose. Are there many more women like your wife?"

"Why don't you ask her?" said Major Love.

A few days later, Nancy Love met with Colonel Tunner. There were hundreds of women, she assured him, who were proven, capable pilots. Not knowing of Cochran's report the summer before, or of Arnold's command to General Olds in January, Tunner launched into an enthusiastic discussion with Love about a female flying corps to serve the ferrying division's domestic wing. Love knew her opportunity had at last arrived.

When Love left his office, Tunner wrote a memo to General George. His answer came back immediately. On June 11, General George told the AAF Chief of Air Staff he wanted to hire women, and Nancy Love was to be transferred to Washington to help Tunner come up with a complete proposal.

Within a week, Nancy Love once again drafted a proposal for hiring women ferry pilots, including recruitment qualifications, what orientation they would need, and how women should be utilized in the ferrying division. On June 18, Tunner submitted her plan to General George.

Evidently General George was so pleased with the prospect of this untapped source of pilot potential that toward the end of June, he mentioned his intention to hire women ferry pilots to General Arnold. In response, Arnold mused that he thought he would talk with the President about it. After all the national interest in women pilots flying for the government, FDR might wish to make any announcement himself.

But Nancy Love's proposal ran into its own snags. Former Ferry Command head Robert Olds, theoretically, had wanted to hire women on the same basis as men—that is, with the same qualifications, at the same salary, and with the same orientation program and initial duties. This surely would have been the simplest procedure. Now that he was gone, however, thinking in the ferrying division had shifted, and there were more officers Love had to satisfy who were not as sanguine as Olds. Love's plan of June 18 to General George reflected her first of many compromises. Though men ferry pilots recruited from civilian life could be between nineteen and forty-five, women would have to be between twenty-one and thirty-five years of age. The women had to have a high school degree, though men needed to have completed only three years (the average grade completed for men and women in 1940 was the tenth grade). And though a month before, the ferrying division had reduced the required number of hours for men pilots from 300 to 200 in response to its shortage of recruits, Nancy Love's women pilots would have to present 500 hours of flying experience. And yet, in spite of this substantial discrepancy,

Love had to restrict women to flying the AAF's smallest trainer and liaison airplanes.

Nor would women be completely integrated into the system. It was decided that they would first all be based together, in effect as a separate squadron, on one ferry base, the Second Ferrying Group headquarters at New Castle, Delaware. The choice was a practical one. Manufacturers of primary trainer and liaison planes were nearby.

So far, though more complicated than it needed to be, Love's plan was moving ahead smoothly. As Cochran had suggested the summer before, Love, General George and Colonel Tunner assumed that women pilots would be hired as civilians, just as men were, and that after a ninety-day trial period, would be commissioned into the AAF. By June 1942, the AAF had an official women's auxiliary, the month-old WAAC. Love and Tunner contacted WAAC head, Oveta Culp Hobby, who heartily approved of commissioning women pilots into her organization.

But in early July, Love discovered that Congress, in its authorizing legislation establishing the WAAC, had made no provision either for flying officers or for flight pay. Women pilots could not be legally commissioned in the WAAC until the legislation was amended.

This was a blow. If they could not be WAACs, no one knew how to pay women pilots flying for the AAF. Getting an amendment through Congress might take months, and the ferrying division was desperate for pilots immediately. Tunner and Love decided to risk an indeterminate future for the program—women would be hired as provisional Civil Service employees. A bill amending the WAAC legislation would be introduced as soon as possible to face its fate in Congress.

On July 13, a new proposal reflecting their solution went to General George, the head of the Air Transport Command. Though men civilian pilots earned $380 a month, the women's salaries were pegged at $250. As if their women pilots' noncommissioned civilian status made the program harder to sell, Tunner and Love increased the recruitment requirements. Now each woman had to have a 200-horsepower rating (planes of that size rented for $40 to $60 an hour), a requirement not imposed on men until the following November. The women also had to supply two letters of recommendation. At the same time, the proposal reassured General George that women ferry pilots would fly only small trainer and liaison-type airplanes.

Five days later, on July 18, General George submitted the ferrying division's proposal to the AAF Commanding General. George called for a small group of highly experienced civilian women pilots for the purpose of "determining the suitability of utilizing women pilots in the delivery of military aircraft." The women ferry pilots would be called the Women's Auxiliary Ferry Troop, or WAFT. This final proposal inevitably included two more requirements: United States citizenship and a mandatory thirty-to-forty-five-day orientation period of flight training and ground school before the WAFTs would go on actual duty as ferry pilots.

The name, General George lamented in his covering letter to General Arnold, did not have quite the psychological appeal of the WRENS (the British Admiralty's new women's auxiliary), but it could be "easily popularized." As for the troop's civilian status, the AAF could "go to bat" for them later in Congress.

For twelve days, as the newspapers reported daily German advances deeper into the Soviet Union, and the heavy midsummer heat hung over Washington, Nancy Love and her Air Transport Command colleagues waited for General Arnold's response. After their detailed preparation, they could think of no reason Arnold would reject the proposal, save one—Jacqueline Cochran.

At last, on July 30, General Arnold responded. He sent the entire proposal back to General George and directed him to confer with the Civil Aeronautics Administration and the Civil Air Patrol and provide Arnold with up-to-date statistics on the availability of women pilots. The reason for this request is not clear, but it appears that Arnold was balking. A few days later, he left Washington—for England, where Jacqueline Cochran was anticipating the arrival, at the end of the month, of her last four American women pilot recruits to the British Air Transport Auxiliary. As of July 4, when the first American crews had flown in an RAF raid on airfields in the Netherlands, American air operations had begun over Europe and future plans demanded Arnold's attention. But while in Britain, he would also meet with Jacqueline Cochran in London.

August was a long month for Nancy Love. She dutifully gathered the requested statistics and the proposal sat ready to be resubmitted as soon as Arnold returned from Britain. But assuming a conference on her plan between Arnold and Cochran inevitable, she was less than confident about the fate of her program.

On September 3, General George gave the proposal back to General Arnold. But this time, George indicated that the hiring of women pilots was ready to be implemented in twenty-four hours, as if telling General Arnold all he needed to do was lift an eyebrow or clear his throat. On September 5, General Arnold did something that was interpreted by General George as activating the Women's

Auxiliary Ferrying Squadron or WAFS (the Troop was now, more sonorously, a Squadron) and the Air Transport Command's ferrying division named twenty-eight-year-old Nancy Harkness Love as director of its new women pilots group. That day, her first telegrams went out from New Castle Army Air Base to her list of experienced women pilots who had been waiting so long to serve their country.

Also on September 5, in England, Jacqueline Cochran bid farewell to her final group of ATA recruits, who were settled into their first phase of training at the Luton ATA base outside of London. General Arnold had told Cochran that the time had at last arrived for her to organize women pilots on the home front, and her air passage was booked for New York.

The public announcement of the formation of the Women's Auxiliary Ferrying Squadron was set for Thursday, September 10. General George and Nancy Love were invited to General Arnold's office that morning to join the AAF Commanding General when he met with the press. When they arrived on Thursday morning, however, they were told that General Arnold had unexpectedly been called out of Washington. Instead, the public announcement would be made by Secretary of War Henry L. Stimson, to whose office they were to report immediately.

The evening papers on September 10 carried the long-awaited announcement by the Roosevelt administration of the first women pilots group to fly for the armed forces. Secretary of War Stimson enumerated the minimum requirements and told qualified applicants to write to the Air Transport Command, Army Air Forces, War Department, Washington, D.C. Immediately, on stepping off her airliner that evening, after the long transatlantic trip from Britain, Jacqueline Cochran picked up a newspaper and was hit with this news. As she recalled in her autobiography twelve years later, she was "mystified."

Friday morning, General Arnold was back from his out-of-town trip to keep an appointment with Cochran. He may have wished he had not returned. Cochran stormed into Arnold's office with the press clipping in her hand. "What's this all about?" she demanded. If Arnold was not as angry as she, he made a convincing show. "He was mad all over," Cochran later wrote, "and when mad, General Arnold could make the fur fly." He had asked the Air Transport Command to prepare plans, he told Cochran, but he had expected to be able to submit those plans to her for study and approval. Instead, the Command had gone over his head and right to the Secretary of War with the project. Arnold picked up the phone and called

Air Transport Command head, General George, and amends were quickly made. Though Cochran could not have the whole women pilots' program, she could at least have half.

On September 15, the War Department announced the formation of another AAF women pilots group, a flight training program to prepare women pilots to serve with the Women's Auxiliary Ferrying Squadron. This second program would be called the Women's Flying Training Detachment (WFTD) and director of women's flight training was Jacqueline Cochran. The following day, September 16, Jacqueline Cochran went to work with the AAF Director of Individual Training as a dollar-a-year woman.

On September 1, 1942, Eleanor Roosevelt, in her famous syndicated column, "My Day," offered that "women pilots are a weapon waiting to be used." Two weeks later, American women pilots had gone from no government flying organization to join, to two separate programs—one for highly experienced professional pilots to serve with the Air Transport Command ferrying division, and one for women with fewer hours of flying experience to receive military flight training. General Arnold had changed his mind about women pilots.

Nineteen forty-two was proving to be a frightening year for the AAF Commanding General; in fact, it would be America's worst of the war. The Allies were losing on all fronts—Europe, Africa and the Pacific. Before the country could be mobilized to the astounding extent it would be the following year, manpower shortages were claimed everywhere, and especially in the air forces, which demanded so many months of rigorous training to produce competent combat pilots. A man unfettered by decisions of the past, Arnold was quick to adapt his opinions. The time to use women pilots had arrived. For the great military strategist and pragmatic commanding general, a need had to be filled; as for the politics of the venture, they would sift themselves out later. But the AAF was no doubt astonished that two women would have enough determination and clout to get their own women pilots organizations off the ground.

As each had to propose her program as "experimental," proof that women could fly was inherently a factor. Nancy Love was forced to set stringent acceptance qualifications for the WAFS, much higher than those for men.

Jacqueline Cochran, on the other hand, made proof one of the main elements of her program. She promised to select women who met the toughest AAF male physical and intellectual standards, and were of high character (she would interview them herself); then she would guarantee that her women pilots would be trained as strenuously and thoroughly as the AAF's male cadets.

In her proposal of July 1941, Cochran told General Arnold she would need 200 women pilots to deliver the 850 AAF trainer aircraft each month to achieve the delivery of all 12,000 on order by the AAF before December 1942. It was now September 1942, and those numbers had become childs' play. The ferrying division was six weeks behind on delivery orders, which numbered almost 3,000 airplanes a month. Nancy Love's elite corps of WAFS certainly would help. But there was only one way to put more than a handful of women on flying duty—institute an official training program for women as part of the war machine. This was Jacqueline Cochran's approach. General Arnold had no reason to reject Love's proposal. But he also accepted Cochran's idea, which was broader and more far-reaching. Whatever Cochran's superior position of political influence, either through her New Dealer husband, Floyd Odlum, whom President Roosevelt had named head of the Defense Contracts Distribution Division of the Office of Production Management in September 1941, or through her own reputation in aviation circles, she had, over many months, convinced a skeptical AAF commanding general that using women pilots was a good idea, and in numbers. To sell a program larger in scale than a small experienced squadron of WAFS, which is what Love wanted, Cochran argued consistently that women were useful to any extent that they could release male pilots from noncombat assignments. Or in her words, "to decrease the Air Forces' total demands for the physically and mentally perfect young men—the cream of the over-all manpower pool, from which all the Services and industry must draw the men to lead the way to victory." Jacqueline Cochran spoke the AAF's language.

But along with the numbers, Cochran also talked familiarly about military discipline for her women pilots. Love had recommended commissioning the WAFS in the WAAC only as a way to pay them. To Cochran, militarization was far more important a concept. Surely she wanted to command a substantial and prestigious military organization. But militarization meant Cochran could monitor and protect her experiment. While Nancy Love preferred to ignore public attitudes, Cochran well knew that she or any director of women pilots would be perceived both by the Army and by the American public as attracting young girls away from their families and setting them loose in their own airplanes to fly around Army air bases. So she ostentatiously placed herself *in loco parentis*. Only by putting women pilots "under direction and discipline in non-operational as well as operational activities," she wrote in one report, "only by having them carefully selected, assigned, checked and supervised," would she reassure the nation and the AAF, as well as guarantee that "her girls" would be a credit to her and to their sex.

Evidently, Nancy Love and Jacqueline Cochran had widely different personal ambitions; in fact, their understanding and uses of power were so divergent that they were able to work together, respectfully, for two and a half turbulent years. Consistent with her comfortable, privileged background, Love, bright and tenacious as she was, merely sought to join and influence a rapidly growing organization—the Air Transport Command. Cochran, in contrast, and in total consistency with her entire career, insisted on founding an entirely new military program over which she would have top control.

But beyond setting herself and esteemed colleagues up in an official flying unit, Nancy Love was not interested in wielding vast administrative power. In fact, she got out of her ferrying division offices and into the cockpit as often as she could. Furthermore, no matter how well they proved themselves as ferry pilots, Nancy Love's WAFs would have remained an elite select group throughout the war. Even if their 500 hours flying experience were reduced to the 300 hours requirement for men, women of the right age with that much civilian flying experience in 1942 were a privileged and determined few: under 100—twenty-five of whom, of course, were already flying in England. As civil aviation in America was severely restricted during the war, without Cochran's AAF training program, no more women would work up toward WAFS credentials. Once all such qualified women joined the WAFS, Love's program had reached its limits. This would have been fine with Nancy Love. She had offered the AAF ferrying division one solution to its chronic pilot shortage and credit to her would be duly recognized by the male high command—so long as women did not make demands considered inordinate. Such a thing would be unbecoming. And she never would. But, in fact, she had everything she wanted.

Cochran, on the other hand, made demands from the very beginning. That she did not always get what she demanded did not inhibit her asking. And this included the directorship of all women pilots, in training and on duty, which she would demand and win in June 1943. But she thought that a corps of women, directed by a woman, would not get lost in an intransigent military hierarchy, which by that time had proved itself slow to allow women to fly more complicated equipment and more dangerous missions than just ferrying trainers and liaison airplanes. Once it was clear that for women proof in operational flying was not enough, Cochran wanted to trouble shoot, to forge new assignments for women outside the Air Transport Command. She also wanted to make sure that once women were trained and assigned, the military system worked for and not against them. She had fought too hard in her own life to trust the military merit system to be entirely fair to women pilots. Jacqueline Cochran may have been impressed by generals, but she certainly was not afraid of them.

In September and October 1942, as Nancy Love's elite and experienced squadron of twenty-eight WAFs, averaging an awesome 1,100 hours of flying experience, were having flight checks at New Castle Air Base, Wilmington, Delaware, Jacqueline Cochran was yet to have a runway for her trainees to land on. But Nancy Love's group of highly professional young pilots were already finding themselves suspicious strangers in the midst of a thriving, and entirely male world of an American wartime air base.

VIII

Above the Crowd: An Elite Corps Is Born

A GUARDHOUSE MP at the entrance to New Castle Army Air Base near Wilmington, Delaware, watched admiringly as a tall, attractive blonde got out of a taxicab. She was wearing a brown hat and brown herringbone tweed suit, a demurely fashionable costume in October of 1942. But as she expertly juggled her three suitcases out of the cab and carried them toward the guardhouse, the MP shook his head in amazement and picked up the guardhouse telephone. "Mrs. Love, here's another one," he said.

A few minutes later, a GI drove up in a jeep. Before he had set both feet on the ground, the young woman had thrown her suitcases into the back seat, hopped in beside him and was fixing on him the most dazzling smile he had seen. "I'm Nancy Batson, from Birmingham, Alabama," she said, "and I've come to join the WAFS."

After a bouncing ride along a rutted dirt road—the entire base seemed to be under construction—the GI pulled up in front of a gray wooden building and escorted Nancy Batson into a drab office containing a filing cabinet, a chair and a desk. Behind the desk sat a young woman with prematurely gray hair, a kittenlike face and wide-set gray eyes fringed with long black lashes. She was Nancy Love, chief woman pilot and director of the Women's Auxiliary Ferrying Squadron of the United States Air Transport Command.

Nancy Batson introduced herself and put her pilot's logbook on the desk. A flawlessly manicured hand, with a huge diamond on one finger, picked it up. When the head of the WAFS saw 650 flying hours tallied on the most recent page, she smiled and got up from the desk. "Excuse me a moment," she said and walked through a door into another office where a man with a black mustache was seated at a desk. On the open office door was the name, "Col. Robert M. Baker," the base commanding officer.

Nancy Batson sat down in the single chair and folded her hands in her lap to hide her agitation. An article in the newspaper on September 11 had told her of the formation of the WAFS. She had simply packed up and taken a train all the way from Birmingham. She never thought of wiring ahead of her arrival, she was so sure she would be accepted. Nancy had all the credentials and, after all, this was a *women's* program. Her luck with men's so far had not been good.

When she got her commercial pilot's license that spring of 1942, she flew her J3 Piper Cub 200 miles up to Nashville, Tennessee, just as her men flying friends had done, to report to Air Transport Command ferrying division headquarters. But she was told that women could not fly for the Air Transport Command, and returned despondently to Birmingham.

Soon afterward, she got her instructor's rating. Exuberant from her check ride, she approached the director of her flying school at the Birmingham airport. She had been a student there for over two years, and one of the school's most dilligent. "I'm now a bona fide instructor," she said, beaming. "I'd like a job. There's nowhere I'd rather work than here, my home base." The flying director looked startled, then mumbled that they were not looking for any flying instructors just then. Nancy was again disappointed. But soon a friend recruited her to come to Miami to teach Army primary cadets at Embry-Riddle Aeronautical Institute, one of the best flying schools in the country. She had been in Miami for two weeks when she heard from home that her old flying school had just hired two new flight instructors—both young men.

Embry-Riddle, however, had given her confidence—she was one of three women flight instructors there—and now that women were being allowed to fly for the Air Transport Command, Nancy was sure she would not be turned away again.

As she sat waiting in Nancy Love's office, a young woman about her age—twenty-two—strode in, wearing zip-up khaki flying overalls. She picked up a pile of letters, glanced casually at the young woman in the brown hat, and confidently strode out. "There's one," Nancy thought admiringly.

After several more minutes, Nancy Love came back into the room. "You may spend the night in the BOQ, where the other girls are staying," she said. "Tomorrow we'll give you a flight test." The two women shook hands and another GI showed Nancy to a long two-story building that looked like a railroad car derailed in a muddy field. To get to the screen door, she had to balance along a pathway of wooden planks.

Inside, she climbed the stairs to a room furnished with an Army cot, a dresser, a mirror and a rod for hanging her clothes. She sat down on the cot. Many minutes later, a tall brunette stuck her head in the door. She introduced herself as Cornelia Fort, from Nashville, Tennessee, and asked Nancy if she would like a cocktail. Upon hearing another southern accent, Nancy brightened, and though in Birmingham a lady did not drink alcohol, she accepted Cornelia's invitation. Once in Cornelia's room down the hall, Nancy eyed the liquor bottles along the dresser top and felt sophisticated. "Bourbon," she said, nonchalantly.

As they chatted, Nancy learned that Cornelia, who was twenty-four, had been a flight instructor in Hawaii. She and a student had watched from the cockpit of a small Cub trainer as the first wave of Japanese Zeros bombed Pearl Harbor.

After their cocktail, Cornelia took Nancy to the New Castle air base officers' club for dinner. Nancy had never seen so many men in one room. (The air base had over 10,000 officers, enlisted personnel and civilian pilots stationed there.) But one table was filled only with young women. "That's the WAFS table," Cornelia said. "We aren't allowed to socialize with the men pilots until we're through our trial period." "Why not?" Nancy asked. "Who knows?" said Cornelia, "but we're following all the rules."

At the table, Nancy met Barbara Erickson, the confident young WAF who had picked up her mail in Nancy Love's office that afternoon. Two weeks before, Nancy learned, Barbara had not been feeling so jaunty. An experienced flight instructor in Seattle, Barbara had received a telegram from Nancy Love requesting that she report to New Castle for a flight test to qualify to fly for the Army Air Forces Air Transport Command. Barbara was so excited that she transferred her CPT students to another instructor in the middle of their course, took out all of her savings and bought a one-way train ticket east.

Two weeks before Nancy Batson had left Birmingham, Barbara Erickson had spent her first night in New Castle in an almost bare room in a BOQ, which had been provided until the WAFS barracks were ready. In spite of Love's detailed planning throughout the summer before her proposal was activated by General Arnold, New Castle was still preparing for the WAFs when the first of them arrived in late September. The women's rooms were still being specially outfitted with mirrors, dressing tables and venetian blinds.

FROM THE AUTHOR'S COLLECTION

The British Air Transport Auxiliary's 15th Ferry Pool at Hamble gathered around a Spitfire. Emily Chapin is standing sixth from left.

ALL PHOTOGRAPHS ARE COURTESY
OF THE USAF MUSEUM,
EXCEPT WHERE OTHERWISE NOTED.

Antiaircraft artillery indoctrination at Camp Davis, North Carolina, where WASPs flew the beachhead towing targets for gunners in training for overseas.

Trainees from 43–5 reassemble an engine during ground school class at Avenger Field.

FROM THE AUTHOR'S COLLECTION

Frances Green, Ann Waldner and Pat Bowser learn to use the top-secret Norden bombsite, equipment in the B–17s they were flying at Lockbourne Army Air Base, Columbus, Ohio.

An energetic moment from the WASP trainees' daily hour of calisthenics, at Avenger Field.

WASP trainee Nelle Carmody, who served as official bugler during flight training, displays her repertoire ranging beyond "Reveille" in an informal concert at Avenger Field.

Avenger Field trainees plotting a cross-country flight.

Trainees study their map before a cross-country trip out of Avenger Field.

Graduation at Avenger Field, Sweetwater, Texas.

Jacqueline Cochran pins AAF wings on Iris Critchell, during graduation ceremonies for the class of 43–W–2, held at Avenger Field, Sweetwater, Texas.

FT, WASP Elsie Dyer and an AAF male pilot at Camp Davis, the wing of an A–24 dive bomber discussing a tow-target ission.

LOW, WASPs at Camp Davis, North Carolina, head for their -24s and A–25s for a day of towing targets and simulated afing, to train antiaircraft gunners.

LEFT, *Barbara Erickson, WASP squadron commander, and Betty Tackaberry marching with the color guard at their Air Transport Command base at Long Beach, California.*

B–17 pilots Frances Green, Peg Kirchner, Ann Waldner and Blanche Osborne after a flight in Flying Fortress, the Pistol Packin' Mama *at Lockbourne Army Air Base.*

Elisabeth L. Gardner, pilot in command of a B–26, at Harlingen Army Air Base, Texas.

COURTESY OF GEORGE ABBATE

WASP P–47 pilots receive their delivery orders of new pursuits in the operations office at Republic Aircraft, Farmingdale, Long Island, in October 1944.

The WAFs, in Hagerstown, Maryland, picking up their first PT–19s to ferry for the AAF in November 1942.

FROM THE AUTHOR'S COLLECTION

Virginia Archer plotting a course in the navigator's compartment of a B–17 Flying Fortress during a cross-country flight.

COURTESY OF NANCY BATSON CREWS

ABOVE, *Nancy Batson climbs into the cockpit of a P–47.* BELOW, *Frances Green grips the throttles of a B–17 as instructor reads the pre-takeoff check list.*

FROM THE AUTHOR'S COLLECTION

COURTESY OF DEV O'NEILL

The WASPs return to Washington, September 1977. President of the Order of Fifinella, Bernice Falk Haydu (RIGHT) *in WASP uniform from 1944, surrounded by WASPs from around the country, presents 25,000 signatures on petitions to Congresswoman Lindy Boggs of Louisiana, WASP bill sponsor in the House of Representatives, on the Capitol steps.*

At about nine o'clock, as Barbara lay on her cot reading, the electricity went out. For an hour, she sat alone in the darkness until, exhausted from her long train trip, she finally fell asleep.

To entertain Nancy Batson at her first dinner at the New Castle officers' club the other WAFs at her table told her stories of arriving at the base and taking their check rides. Barbara Poole from New Jersey, who had become the youngest licensed pilot in America in 1936 when she earned her license at sixteen, and the youngest commercial pilot the following year, had been an instructor with the CPT program. Every one of her students had passed his government tests for a license, a record which had made her unpopular with some of her male instructor colleagues. When she arrived at New Castle, her ATC check pilot turned out to be one of them. After Barbara had executed some neat spins and snap rolls (she had taught them), a flock of geese, heading south down the Chesapeake Bay for the winter, flew by in front of their airplane. Without warning, the check pilot took over the stick and, full-throttle, began to chase one of the geese all over the sky. Barbara was horrified. She had heard that if a goose became separated from its flock, it could never rejoin it. When they finally landed, the check pilot walked away without a word. Not until hours later did she learn that she had passed.

Listening to such episodes, Nancy Batson grew apprehensive about her own flight check. The WAFs, she realized, had all made aviation their profession. Evelyn Sharp, a former barnstormer from Nebraska, had almost 3,000 hours of flying experience. That was the equivalent of flying a light plane 120 times back and forth across the country. In fact, the average number of air hours among the WAFs was 1,162, or the equivalent of making around fifty transcontinental trips, before they joined the ferrying division. One had owned her own airport in New Jersey, another had run a flight school near Los Angeles. The majority, like Nancy, had been flight instructors.

The WAFs' family backgrounds ranged from an orphan to heiresses (Woolworth stores and Luden Cough Drops), and they spanned half a generation in age, from twenty-one to thirty-four years old. Among the more senior WAFs was Betty Huyler Gillies, who had been inspired to learn to fly in 1928 by Amelia Earhart's *Cosmopolitan* article, "Try Flying Yourself," and with Earhart had founded the Ninety-Nines. Betty Gillies was married and had two elementary-school-age children.

After the meal was finished, they lingered over coffee. Nancy Batson was struck by how much she liked these WAFs. In fact, for the first time in her life, she was surrounded by a group of women who shared the same enthusiasm for flying that had so dominated her own life over the past two years—women from Pennsylvania, California, Texas, Missouri, who had also pursued their love of flying alone, as "the only girl." Now they had all been brought together in one place, to comprise a special experiment.

The next morning, Nancy Batson reported to the flightline. She was put through an hour of flight maneuvers—stalls, spins, chandelles and lazy eights, by a taciturn lieutenant. Back on the ground, Nancy felt confident she had done well, but, saying nothing, the lieutenant grabbed her logbook and stalked off toward the administration building. Suddenly the door opened and Nancy Love emerged. "We had a meeting of the ferrying division civilian pilot selection board," Love said. "For the WAFs, I am allowed to sit in on the board reviews as an unofficial member. Don't worry. You're in, you're accepted." Nancy Batson let a long slow smile spread over her face as Love took her to the equipment office. The new WAF was issued khaki flight overalls, a pair of goggles, a parachute and an Army Air Forces white flying scarf.

AVERAGING almost 1,200 hours of flying experience, the WAFs were some of America's most experienced young pilots, male or female, in 1942. When they converged on New Castle Army Air Base headquarters of the 2nd Ferrying Group of the Air Transport Command, the ATC was six weeks behind on its deliveries of new aircraft from factories to America's air bases. Yet when they appeared, at their own expense, to offer their flying skills to the U.S. Army Air Forces, the WAFs were made to feel like fledglings.

Gertrude Meserve, a twenty-three-year-old WAF from Boston, had taught hundreds of MIT and Harvard students to fly as an instructor with the CPT program. But when she showed her logbook of over 2,000 hours to her Ferry Command check pilot, he looked at it critically. "I can always tell," he said, "once I see how a girl flies, whether or not she has padded her logbook."

Though proving their flying proficiency was not a problem—the WAFs had more flying experience than some of their Ferry Command instructors—they were still put through a special forty-day orientation period before being allowed to fly the Army's smallest airplanes. Thus, the highly experienced, 200-horsepower rated WAFs, whose required credentials so overqualified them for entry level civilian jobs with the ferrying division, spent their first six weeks being instructed in aircraft they all had been flying for years. When not flying, they were in intensive ground school, which, aside from Air Transport Command procedures and AAF regulations, many of them had taught.

But as they practiced takeoffs and landings in the PT–19s, at a small grass airstrip formerly owned by DuPont near their air base, the WAFs did not complain. They would rather be flying than anything else, even if they were not allowed to practice the barnstormers' routines which had paid the daily bread of at least three of them. Though their salaries had been much higher as flight instructors than their $250 a month as WAFs, they felt there was a big difference between teaching college students who would soon be pilots and officers in the AAF and flying AAF airplanes as military pilots themselves.

In the fall of 1942, as the WAFs reported to New Castle, the war in Africa was in full force, and Americans would soon be fighting there in large numbers. American Pacific installations on Guadalcanal in the Solomon Islands were under Japanese attack. Isolationist sentiments had long ago evaporated. At last, to Americans, the war seemed very real. And given the opportunity to fly airplanes for the war effort, the WAFs would not have traded their places in PT cockpits with any women in the world.

Every morning when Gertrude Meserve climbed into the open cockpit and looked out at the white star in the big blue circle on her wing, she was filled with pride. While flying at Logan Airport in Boston, as a student in 1938 and, later, as an instructor, Gertrude gazed longingly at Army P–40s, then the hottest military pursuit planes, which flew in periodically to refuel. To her, they were Oz with wheels and a propeller. But she had thought that only men flew military airplanes—the young men she was teaching to fly. Yet here she was.

Flying military airplanes, in which cockpits were built for men, presented a challenge to WAF Betty Huyler Gillies, who was five feet one and a half inches tall. When she was a utility pilot for Grumman Aircraft, on Long Island, she could adjust her seats in the small liaison planes. But in some Army planes, even with a couple of pillows behind her back, Betty could not reach the rudder pedals. She then recalled a young Grumman test pilot who was about her size. Telephoning him on Long Island, Betty learned that he made special wooden blocks which fit over the rudder pedals so his feet could reach. Betty asked if he would make her some. The next time her husband, Bud, a Grumman executive who often had business with the Navy Department in Washington, flew through Wilmington, he bought his wife a pair of shiny, carved wooden rudder pedal extenders.

During their trial period, the WAFs accepted their apprenticeship with good humor, not only because flying for free would always be worth the repetition, but because they were keenly aware that their every move was being watched by headquarters, that they were an experiment, and not an auspiciously anticipated one. No WAF could ignore the heat of notoriety.

Whether or not the WAFs were viewed as having any potential permanent value, they were immediately perceived as highly vulnerable to scandal. Long before the WAFs had finished their orientation, the ferrying division was already worried about how they were going to function within established ferry operations. The first directive from headquarters delineating rules and regulations reached the WAFs on October 12. "Under no circumstances," they were warned, were they to solicit rides in bomber-type airplanes for local or cross-country flying. Nor were they to try to secure rides from an Army air base back to home base after a ferrying delivery. This regulation, though meant to obviate any "Airport Annie" charges against the WAFs, in practice deprived them of a vital means of transportation available to men pilots, and caused all women ferry pilots over the next two years considerable and unnecessary time and fatigue. Without the option to hitch rides, they had to get back to home base either by overcrowded trains and buses, or on commercial airlines, which landed only in major cities (most PT deliveries went to out-of-the-way training bases) and followed skeletal wartime schedules. Later, when they were flying larger airplanes, they were forbidden to fly ferry trips in the same cockpit—even in the same group of planes—as men pilots. Nor could they carry men passengers, though they could fly one another as passengers and could ride as passengers of a male pilot. Eventually, some of these restrictions were ignored by base commanders as the WAFs amply demonstrated their morals as well as their metal.

One restriction, however, met with unanimous passive resistance. In March 1943, the Air Transport Command ordered its women pilots grounded for the week they were menstruating. This command was consistent with the federal government's thinking elsewhere. The 1942 Civilian Aeronautics Administration *Handbook for Medical Examiners*, which directed the doctors who conducted physicals required for a student pilot to solo, stated: "A history should be obtained of any menstrual abnormalities, pregnancies and miscarriages. All women should be cautioned that it is dangerous for them to fly within a period of three days prior to and three days after the menstrual period. . . ." In a matter of weeks, the ferrying division realized that the order was not being obeyed. But if Nancy Love was not enforcing it, none of the men in high command was particularly willing to try monitoring women pilots in this regard. The regulation was formally dropped. The WAFs had flown almost daily from December through March without a single accident, not even a minor mishap.

By the end of October 1942, the first WAFs to report to New Castle had at last finished their forty-day orientation period, imposed on them by the ferrying division. "Like the temptation of Christ in the desert," moaned WAF Teresa James. She and several others held their first actual delivery orders in their hands. They were to pick up L4 Grasshoppers, 65-horsepower observation/scout and liaison airplanes manufactured by Piper Aircraft in Lockhaven, Pennsylvania, and fly them to air bases in western Pennsylvania. The L4, the Army designation for Piper Cub, cruised at about seventy-five miles an hour, which with a headwind meant a ground speed slower than the traffic on the highways below. But in their delight finally to be on duty as legitimate ferry pilots, as well as to be traveling the airways as they had dreamed of doing for so long, the WAFs did not care how small or slow the plane. They piled into a Lockheed Lodestar (the successor model to Amelia Earhart's Electra) and sang exuberantly all the way to Lockhaven.

When they arrived, they found the Lockhaven airport busy with young women assembly line workers practicing takeoffs and landings in Piper Cub trainers. The Piper workers greeted the WAFs with awe and excitement. "Come on and get your licenses!" the WAFs told them. "Come fly with us!" Every time the WAFs came in, the Lockhaven workers would fly harder, knowing that they soon would be able to earn a living flying airplanes rather than fabricating them.

By the time Nancy Batson got her first ferrying assignment, November had arrived and the WAFs were being assigned to deliver their first PT–19s, manufactured by the Fairchild Engine and Airplane Corporation factory in Hagerstown, Maryland. Unlike the L4s, which had cabins, the PT–19s had open cockpits. On November 7, Nancy and five other WAFs received their first delivery orders: the planes were to be delivered to a training base in Chattanooga, Tennessee, a 650-mile flight from Hagerstown. They all reported excitedly to operations at New Castle and were issued Army winter flying gear—bulky fleece-lined leather jackets, high-waisted fleece-lined leather pants which zipped from the shinbone of one leg up to the sternum and were held up by suspenders; gloves; chin-strapped leather flying caps; wool-lined boots, and for under everything else, woolen long johns. "You think this is a lot, but it won't be enough," said the equipment officer. A couple of the WAFs refused to wear the flying pants.

As they all waddled to the transport which would fly them on the hour's trip to Hagerstown, Nancy joked that she could roll down the ramp more easily than maneuvering upright. Everyone's spirits were high.

Once they arrived in Hagerstown, in the western Maryland mountains, they had to wait half the morning before a grim November weather ceiling lifted high enough off the mountains for them to take off. Finally, they walked out to their shiny new PT–19s, fresh off the Fairchild assembly line and never flown before, climbed into the cockpits, and took off southwest toward Chattanooga.

As she ascended to 3,000 feet above the Appalachian Mountains, the belle of Birmingham began to realize how unfriendly the northern sky could be. The air was flowing past the open cockpit at over a hundred miles an hour and the temperature hovered around ten degrees. Nancy had to lash her charts to her leg so they would not be blown away. PTs had no radios, but since she could see the other little primary trainers dotting the sky around her, she knew she would not get lost. Every two and a half hours, they would land to warm up and refuel, and to relieve themselves of the hot coffee gulped gratefully during the previous stop.

The flight took almost eight hours. When they landed in Chattanooga, the sky was almost dark. The WAFs staggered to a bus reserved to drive them into town where hotel rooms awaited them. Nowhere was a restaurant open for dinner. Frozen and exhausted, they went silently to their rooms. Nancy sat on her bed, fighting back tears, her teeth still chattering and her fingers too numb with cold to tackle all of the zippers of her flying suit. "I don't believe I am going to be able to do this," she confessed to herself miserably.

At breakfast the next morning, after a good night's sleep and a hot shower, she felt more game. She also learned that she was not the only WAF who had suffered second thoughts about her adventurous wartime duty.

WHILE HER WAFs were busy on their first ferrying assignments, Nancy Love was preparing a surprise for them—uniforms. Since she could not turn to the AAF to outfit her civilian flying corps, she searched downtown Wilmington, Delaware, until she found a tailor who would make the official WAF uniform—a gray-green belted jacket, an open-collared light gray shirt, and gray-green slacks—for a reasonable charge. The civilian WAFs would have to buy their uniforms themselves.

When she announced that the WAFs were to go into town to be fitted for uniforms, the news was received with as much excitement as their first ferrying orders. In an era when a uniform made the man, the WAFs felt that they were about to receive the final mark of legitimacy. The afternoon they were ready, Nancy Batson stood tall before the mirror in the tailor's fitting room and then, somewhat self-consciously, stole out of the shop and down the street to a photographer's studio to have her picture taken to send home to Birmingham. Shortly, she was joined by several other WAFs who entered the studio looking embarrassed and then burst into laughter as they saw they were not alone in their irresistible vanity.

Now that the WAFs were uniformed, however, they were ordered by Colonel Baker to march with the men of New Castle in the weekly air base parade review. Their first invitation came just in time for a visit by several AAF generals from Washington on an inspection tour.

The WAFs now knew how to fly the Army Way, but none knew how to march. All week before their debut parade review, they formed two lines of twelve outside their ground school classroom and, under the stern eye of a drill officer, practiced marching. At first, the group felt pretty motley. As the drill officer shouted the command, "Eyes right!" WAF Teresa James threw her head so violently to the side that her cap stayed pointing straight ahead. Then each WAF had to take a turn as drill sergeant and call commands. On one practice drill, a WAF got the little squadron marching ahead at a good clip. She was pleased. The rows were all in line, everyone was in step. Suddenly she realized that her columns were heading straight for a ditch. She began shouting several unintelligible commands, none of which was, "To the rear, March!" The new female precision drill team piled up in a bunch at the edge of the ditch and dispersed in gales of laughter. The drill officer watched the bedlam with a "those silly girls" look on his face.

On Saturday morning, the WAFs spit-polished their shoes, brushed their new uniforms, and spent many anxious minutes powdering their noses and putting on lipstick in front of their mirrors. They all had a hard time placing their hats correctly, so that they looked official but not unflattering. Then they met Nancy Love outside the BOQ. Love, who was to march at their head, asked casually if they knew how to march yet. With energetic nods, they assured her they did. Love laughed. "I don't know how to lead you," she said. "Whatever I don't say, you do it anyway." They told her they had a lot of practice at that.

The New Castle parade ground looked like an Army football game at halftime, as the huge gray-green blocks of officers and enlisted personnel marched up and down in front of the reviewing stand. One by one the squadrons paraded past the line-up of beribboned generals who eyed them sternly or, lost in conversation with another officer on the stand, ignored the spectacle altogether. After half an hour, the parade reached the end, with the tiny two-column squadron of WAFs clad proudly in their new gray-green uniforms marching with alacrity. Nancy Love yelled for them to halt and make a left face, so that the two columns would face forward toward their illustrious guests. As soon as they turned, however, they found that their two marching columns had been too close together. The second row found themselves with their noses almost in the hair of the WAFs in the front row. Afraid to move back, they simply stood at attention.

The generals gazed impassively along the formations of the parade. Then their eyes came to rest on the small female squadron at the end of the line. The WAFs caught their breaths as the entire group of generals left the stand and began to walk toward them. Of all the corps in the parade, they had decided they would review the WAFS more closely. The WAFs in the front row looked straight ahead, expressionless, as the generals walked past them and appraised each WAF from cap to spit-shined shoes. As they got to the end of the row, the WAFs breathed easier, thinking their ordeal was over. But the generals had decided that they would review the second row more closely as well. The WAFs' marching mistake had suddenly become crucial. One by one, the generals side-stepped between the two rows of WAFs, as the more buxom of the squadron tucked in their chins and sucked in their breaths in a hopeless attempt to give the generals room to pass. One WAF and a general almost locked chest decorations. Watching the drama out the corners of their eyes, the young men in the neighboring formation were about to explode.

Finally, the generals filed back to the reviewing stand and the parade was dismissed. It was now noon. As the WAFs, much relieved, headed for the officers' mess, Nancy Love pulled a few of them over to the side. They had been invited to lunch with the generals, she told them. Most had been the uncomfortable young women of the back row. As the more lithe and wiry WAFs walked on toward the WAFs' table they laughed about why *they* had not been selected for luncheon with the generals.

THE WAR DEPARTMENT ceased all publicity on the WAFs just as they began flying across the country to deliver airplanes. As conspicuous as they were at their home base of New Castle, for the rest of America they were an almost unknown government flying squadron. Even in their new uniforms, they were taken for WAACs, Girl Scouts, or even members of the Mexican Army.

As winter had set in, weather was becoming unpredictable, and the WAFs often had to land on air bases en route to their delivery points where the existence of an official corps of women pilots was totally ignored. In the minds of many Army officers, a woman on an air base was there for only one reason.

In early December, Barbara Poole ran into some ominous rain clouds over Georgia while flying a PT–19. Checking her charts, she located the nearest Army base, which was Camp Gordon, and as the sky got darker she flew lower until the field was in sight.

As her PT had no radio, she landed unannounced, and when she walked into the operations office and pulled off her flying cap, she was met by complete consternation. When she asked to file an RON (an AAF acronym for Remain-Over-Night) for weather, the officer behind the desk picked up the telephone. In what seemed like seconds, the highest commanding officer of the base strode into the room. "Young woman, I want you off this base immediately!" he shouted. "I'm staying," Barbara said. "There is a front coming in and it is unsafe to fly." The commanding officer did not believe that she had just landed in the Army trainer parked by the hangar. Again he ordered her to leave the base. Barbara asked to use the phone and called Nancy Love at New Castle Army Air Base. But when the director of women pilots in the Air Transport Command got on the phone, the commanding officer did not believe her, either. "I don't know what kind of joke you girls are playing, but you're not going to play it on my base," he roared and hung up. Barbara was ordered off the base and, as parting words, was warned, ". . . and don't come back!"

Barbara looked out at the PT by the hangar, picked up her parachute and briefcase and walked out the door. At the front gate of the base, she got a bus into town, where she checked into a hotel until the weather lifted, then she sneaked back and took off as fast as she could, vowing never to return to Camp Gordon, Georgia.

As THE WAFs' flying skill overcame their initial stage fright, ferry trips from Hagerstown down the East Coast to the southern states and Texas, where most of the Army primary training fields were located, became routine. They began to imitate the men ferry pilots by flying their PTs in formation, confidently easing their wingtips to within inches of one another. Returning from ferry trips, they laughed at their bizarre sunburns they had acquired in their open cockpits. Depending on whether they were flying south or north all day, one side of their faces would be scarlet, with a goggle-sized white hole around one eye.

They also got used to inclement weather, which plagued them in primary trainers that had no radios or navigational aids. But as far-flung as they were becoming, working together and flying together, friendships between WAFs were stronger than any they had ever known. After spending her first lonely night in the blacked-out BOQ listening to her male neighbors play poker, Barbara Erickson was taken under the wing of WAF Betty Gillies. In typical Army fashion, nicknames quickly emerged. Barbara, who was one of the youngest WAFs, was called "Daughter" and Betty, one of the eldest, became "Mother."

Mother and Daughter flew many cross-country trips together in PT–19s during December, and they inevitably found themselves weathered in all over the East Coast, sometimes for days at a time. Ferry pilots were allowed six dollars per diem while away from home base on trips, but money was no cure for boredom and agitation to get going.

On one ferry trip, Barbara and Betty were grounded in Albany, Georgia, by a winter storm. As movements of aircraft by the ferrying division were top-secret, ferry pilots were required to send a wire to their home base in wartime code each time they chose to RON. The message had to tell where the airplane was and why the pilot was stopping the flight. A common code message, especially for pilots ferrying PT–19s, in which every meteorological disturbance was drastic, was "THREE," which meant grounded by weather. Betty Gillies always sent a second wire to her husband at the Grumman plant on Long Island, letting him know where she was.

For four days, Betty and Barbara had wired "THREE" from Albany, Georgia. They had seen both movies in town and they were getting restless. When they learned from air base weather forecasters that rain was expected for yet another day, Betty and Barbara looked skyward in despair. Then Betty got a gleam in her eye. They wired "THREE" to New Castle, but for the first time Bud Gillies on Long Island was not sent an identical message. Betty wired him, "HEBREWS THIRTEEN EIGHT." Production nearly stopped at the Grumman aircraft factory that afternoon while management tried to decipher Betty Gillies' message. Finally someone ran up to Bud Gillies with a King James Bible open to Hebrews 13:8. It read: "Jesus Christ the same yesterday, and to day, and for ever."

As THE WAFS successfully delivered an increasing number of PT–19s and Piper L4s throughout December and January, the watchful eyes of Air Transport Command headquarters began to see the unexpected. As an experiment, WAF Teresa James was made flight leader —the pilot of the lead plane who is responsible for navigation—on a flight of twenty-three PT–17, open-cockpit Stearman bi-planes, between Great Falls, Montana, and an air base in Tennessee. Six of the Stearmans were to be flown by WAFs, the others by men ferry pilots. The whole group boarded a train in Wilmington for the two-day trip to Great Falls.

After taking off from Great Falls, the flight scattered, each pilot taking his or her own pace for the long cross-country flight. The morning of the second day, the PT–17s began to arrive in Tennessee. One by one, the six WAFs landed, but two days later only six of the seventeen men pilots had appeared. As radio reports filtered in, it became clear that the others had got lost; a couple had damaged airplanes, others were suspiciously late, presumably paying visits along the route.

The Air Transport Command was horrified at the 50 percent safe delivery rate of the flight. But headquarters could not avoid noticing who had accomplished their missions successfully. Nancy Love began to see that her little WAFS, after such a doubtful beginning, just might shame the ATC into shape.

The week before Easter, 1943, the WAFs finally made their point so loudly that even they heard the reaction from high command. WAFs Betty Gillies, Nancy Batson, Helen McGilvery, and Sis Bernheim (both former flight instructors from New York), were ordered to fly four PT–26s, Canadian RAF primary trainers which were PT–19s with canopy-enclosed cockpits, from Hagerstown to Calgary, in Alberta, Canada—a distance of over 2,500 miles. They left Hagerstown on Palm Sunday.

Flying hard all day, they ran out of sunlight just south of Chicago, in Joliet, Illinois. Hotel rooms in wartime Joliet were hard to find. After searching for two hours, they finally checked into a second-rate boarding house. To find a restaurant open for supper took another hour. When they finally got back to the boarding house, it was ten o'clock. They had been up since dawn. Before retiring to her room down the hall, Betty, who was flight leader, said, "Now, we're going to get up at 4:00 in the morning to get an early start." Lying on their beds dead tired, Helen and Sis were grumbling, "We went seven hundred miles today." "I believe I could," drawled Nancy sleepily, "and maybe we should. I've a feeling something's going on here." Then the three went to sleep.

At four, the telephone in their room rang. Nancy fumbled with it until it reached her ear. "Time to get up," said Betty cheerily. Without a word, all three WAFs rose, got dressed in their flying suits and joined Betty in the lobby. When the sun broke on the horizon, four PT–26s sat, engines running, at the start of the runway, and one by one, took off toward the northwest.

That day, they flew another 600 miles, to North Platte, Nebraska. On Tuesday, they arrived in Great Falls, Montana, a large United States ferry base 100 miles from the Canadian border where planes were processed for lend-lease delivery to Canada, and, for aircraft bound for Russia, to Alaska. That day, the WAFs had achieved 850 miles, a prodigious flight in airplanes that averaged a ground speed of about 100 miles an hour. Before finding a hotel in Great Falls, Betty maneuvered their PT–26s through customs, and on Wednesday, the four WAFs were flying along the spectacular snow-capped peaks of the Canadian Rockies on their way to Calgary. Early in the afternoon, they signed the papers of four PT–26s, safely delivered from Hagerstown to Calgary in a record three days, for which their base commander was officially to commend them—and the entire Women's Auxiliary Ferrying Squadron.

In only six months, the WAFs had gone from fledgling pioneers to seasoned ferry pilots. Beginning in January 1943, Nancy Love took small groups of WAFs from New Castle to establish more women pilot squadrons in the three other major ferrying groups across the country: Love Field, Dallas, Texas; Romulus, Michigan, outside of Detroit; and finally, in mid-February, Long Beach, California.

Little by little, the Air Transport Command was easing its original skepticism which restricted women to the cockpits of trainer aircraft.

Nancy Love and the WAFs would soon check out in much bigger airplanes, a story to be told in Chapter 13, and would forge the way for Jacqueline Cochran's training school graduates who would begin to augment the WAFS ranks in May. But as the most experienced, the WAFs would remain the elite—an elite without stripes—of all of the women pilots flying for the Army Air Forces.

IX

Learning to Fly the Army Way

In late September 1942, Jacqueline Cochran leaned over her desk at Army Air Forces headquarters and looked twenty-two-year-old Mary Trotman in the eye. "I want to start my own training school for women pilots," Cochran said. "They won't let me unless I can prove women can fly Army airplanes. I've contacted all of you who have a lot of flying experience and you're going to show them women can do it. The program depends on you. Are you interested?"

Mary Trotman nodded automatically. Her husband, Ed, a counter-intelligence officer, was about to be shipped out to England, so she had no reason to reject an opportunity like this. Four years before, Mary had taken an airplane ride with a military test pilot. The next day she took her first flying lesson, and the day she soloed, she asked her husband if she could buy an airplane. The young Trotman household had been transformed. When Ed came home from work, Mary had just returned from the airport. They began eating supper out.

When Ed joined Army intelligence just after Pearl Harbor, Mary continued flying. Now, nine months later, she had earned her Civil Aeronautics Administration commercial license which qualified her for Jacqueline Cochran's Army Air Forces Women's Flying Training Detachment, the WFTD. She was just getting organized, Cochran told Mary, but within a month she could expect a call informing her when and where to report for duty.

Three years had passed since Jacqueline Cochran had written to Eleanor Roosevelt, in September 1939, suggesting the formation of a women pilots organization. Now she had a beginning. While Mary Trotman was arranging to sublet her Chevy Chase, Maryland, apartment and to store her airplane for the duration of the war, Cochran was interviewing dozens of women pilots who had at least 200 hours of flying experience. They did not have the formidable qualifications demanded by Nancy Love's WAFs, being recruited in New Castle, Delaware, but Cochran wanted to guarantee that her group had the fortitude to stand up under the grueling regimen of an AAF cadet flight training program. The 200-hour flying experience necessary to apply would soon have to be lowered as more classes needed to be filled (in April, hours necessary to apply were dropped to a minimum of thirty-five), but she would not compromise in character.

Nor would she compromise in physical requirements. For Mary Trotman and the other women pilots she interviewed, Cochran arranged at an air base near their homes for a physical examination. Identical to that given to Army Air Force cadets, the physical was designed for prospective combat pilots. It was called the Form 64, and often required an entire day with flight surgeons and in laboratories. Her women had to be tops, like their male counterparts.

Cochran was finding the women she wanted for her training school. But she was also beginning to find that though General Arnold had ordered a school into existence, maneuvering for all the details was involving more military commands than some of the war's most complicated battles.

General Arnold had given his order on September 13 to General George Stratemeyer, his Chief of Air Staff. Arnold suggested that Cochran "pick out, say, the five hundred best women fliers she can find in the U.S." and send some to the WAFS and some to her training program. The very best were already applying to Nancy Love to join the WAFS. As for Cochran's training school, General Stratemeyer organized a conference for her on September 15 with the AAF Director of Individual Training, Colonel Luther S. Smith, and his staff, some Civil Aeronautics Administration civilian pilot training experts, and a young officer, Captain James I. Teague, from the Air Transport Command ferrying division, which would hire training school graduates.

The conference took all afternoon. A CPT-type curriculum quickly emerged. But on the question of flying experience for her training school recruits, Cochran ran into opposition from Captain Teague of the ferrying division. The WAFs were required to have 500 hours of flying experience, and the ferrying division was delighted to hire women of such caliber. "That's ridiculous," Cochran said. "This is a training program. I could take a woman pilot with fifty or seventy-five hours and a private license and give her fifty hours and put her in a P–40 or similar hot plane. I could not find more than a couple hundred women in the whole country with the kind of experience you're asking for." Teague insisted that the ferrying division had to maintain its standards. "All right," said Cochran, "the ferrying division hires men civilian pilots with three hundred hours of flying experience. If I accept women with commercial licenses, they'll have three hundred hours when they come to you." Teague compromised when Cochran added to her training curriculum instrument flying experience and fifty hours in airplanes of over 200 horsepower.

After the conference, however, Captain Teague was uneasy. In a lengthy report to Colonel William Tunner, head of the ferrying division, he expressed his concern that Jacqueline Cochran seemed more interested in expanding her training program than meeting the ferrying division's qualifications. On September 17, Colonel Tunner wrote a special report to General Harold George, assuring the Air Transport Command chief, who had met with Cochran five days earlier upon her return from England, that the ATC ferrying division would be hiring *only* those female training school graduates who met its 300-hour flying experience requirement. Tunner's responsibility was to get airplanes delivered intact, and he already sensed that the AAF was going to dump its women pilots in his lap, whether he wanted them or not. He still viewed Cochran's future training school graduates in the same category as the male civilian pilots he was hiring singly. He was not able to consider Cochran's program as an Army cadet school, from which later he would gratefully accept male graduates with only 200 hours of training. All he could see now was that his bailiwick was threatened by the ambitions of a strong-minded woman.

But as September drew to a close, Cochran was by no means getting her way. Now that her recruitment standards were set, Cochran needed a school. To her dismay, the Civil Aeronautics Administration announced that they were already over-committed to the War Training Service program, and did not have the resources to organize and run a women's flight training school. (Evidently the same thinking ruled as in the summer of 1941, when women college students were banned from CPT courses run by the CAA.) Yet Colonel Smith's Office of Individual Training had no flight training facilities of its own, nor did the Air Transport Command. For several days, Cochran was in limbo; though her women pilots training program was civilian, civil aviation authorities would not take responsibility. Though the AAF had ordered it into existence, the Army would not accept civilian women pilot trainees under its jurisdictional auspices. As a last resort, Cochran went to General Arnold and explained her dilemma. It was evidently quite a meeting.

On October 7, the AAF Commanding General ordered his Flying Training Command, under the leadership of General Barton K. Yount, to organize Cochran's civilian women pilots training program. In his directive to the training command's Fort Worth, Texas, headquarters, General Arnold encouraged General Yount to start the school "at as early a date as practicable"—the AAF was going to train 500 women pilots. Arnold made no mention of a commercial license recruitment requirement for training school applicants; in fact, he suggested seventy-five hours "insofar as possible." "Limitations of hours will not be published," he wrote Yount. "Individuals will be selected upon their qualifications." Jacqueline Cochran had successfully declared her independence of any ferrying division authority over her training program, and General Arnold was giving her his complete support.

It took over two weeks for the Flying Training Command to respond to Arnold's directive. Jacqueline Cochran left Washington for the Cochran Ranch in Indio, California, to wait for the training command's reaction. Finally, on October 27, she received a call from General Yount. He had wired General Arnold that he would accept responsibility for the program, and she was to report immediately to Flying Training Command headquarters in Fort Worth. From then on, Cochran began to see results. In two days, Yount obtained authorization to buy civilian airplanes and to negotiate a contract with a civilian aviation company to run the training school. They found a company that had a CPT program contract that was about to expire. And on November 3, Cochran and Yount designated Aviation Enterprises, Ltd., located at Howard Hughes Field, Houston Municipal Airport, Texas, as the future home of the 319th AAF Flying Training Detachment (Women). The choice of Houston brought Jacqueline Cochran's women pilots training program under the wing of the Gulf Coast Air Forces Training Center, where at last, after almost eight precarious weeks, it found a niche in the vast structure of America's wartime military establishment.

While Nancy Love moved into her office in the New Castle Army Air Base administration building the day after her appointment and began greeting her arriving WAFs, Jacqueline Cochran had to build her organization—one viewed with ambivalence at best—out of a wing and a prayer. And watching how well Cochran could fight her own battles was General Arnold, also feeling pressures of his own to build an air force of pilots for combat in Europe and the Pacific. The day the contract was signed for a women pilots school in Houston, Arnold sent his second directive to the AAF Flying Training Command which indicated that he was now exuberant about the possibilities within the AAF of a women pilots training program and its graduates. The directive was headed "Augmentation of Women's Pilot Training Corps," and it stated:

> Contemplated expansion of the armed forces will tax the nation's manpower. Women must be utilized wherever it is practicable to do so. It is desired that you take immediate and positive action to augment to the maximum possible extent the training of women pilots. The Air Forces objective is to provide at the earliest possible date a sufficient number of women pilots to replace men in every non-combat flying duty in which it is feasible to employ women.

After months of persuasion, Cochran was finally beginning her program with Arnold's adamant support. And apparently, the sky was the limit.

Once she had a contractor in Houston, Cochran immediately hired a staff executive for the Women's Flying Training Detachment—a Fort Worth Red Cross administrator and swimming instructor, Leoti Deaton. Deaton's first assignment was to go to Houston to find housing for the trainees as near as possible to the airport. Cochran told her that twenty-eight women pilots would be arriving in the middle of November, so Deaton had to work fast. As in many American cities in wartime, lodgings were scarce in Houston. After three days combing the city, Deaton finally secured rooms in several neighboring tourist courts, eleven miles from the airport, but they would not be available until December. For the interim two weeks, she resorted to renting rooms in private homes. This was not easy. As the program was designated top-secret, Deaton was under strict orders not to divulge to prospective landladies what the young women would be doing in Houston.

Transportation for the trainees to and from the airport presented the next problem. Not for two or three weeks would the Army supply semitrailer trucks, fitted with wooden boards in the back to serve as seats. Until then, the only mode of mass transportation that the contractor, Aviation Enterprises, could muster was a bus that had been owned by a Tyrolean orchestra. At seven every morning, the twenty-eight participants of the top-secret women pilots training program would be picked up from their boarding houses in a huge white bus outfitted with red and white striped awnings and profusely decorated with edelweiss.

Meanwhile, the Gulf Coast Air Forces Training Center was

trying to find airplanes for the women trainees to fly. Army trainer aircraft were on order, but as at other flight schools, there was a delivery delay in the ferrying division which the WAFs were being hired to alleviate. But by commandeering almost every private airplane between Fort Worth and Houston, the training command finally assembled a flightline for the new women pilots school.

On November 10, telegrams went out from General Yount of the Flying Training Command to the first class of the Women's Flying Training Detachment, Class 43–1, indicating the first graduating class of 1943. All were anxiously waiting for notification. One telegram to a future woman pilot trainee began, "It is desired that you report for duty at your own expense at 10:00 A.M. No provision is made for your sustenance and maintenance during the term of this appointment. No uniform will be issued . . ." But although their invitation was not particularly cordial, twenty-eight women pilots reported on only six days' notice to Howard Hughes Field, Houston, Texas, on Monday, November 16, 1942. Also reporting that day were the base commanding officer, the adjutant and the two Army check pilots. Thus, the 319th Flying Training Detachment (Women) was hurled into existence.

THE TRAINEES recruited for Jacqueline Cochran's Women's Flying Training Detachment in Houston were a different breed from Nancy Love's WAFs, who were delivering their first Army primary trainers back East. Cochran's first class, 43–1, averaged well over the required 200 hours of flying experience—or around 350. Nevertheless, the majority of WFTDs—they began calling themselves "Woofteds"—had flown not for a living, as had Love's WAFs, but because flying was a personal enthusiasm. Reporting to Houston were a Reno blackjack dealer, an actress and the sister of Broadway producer Joshua Logan, a Mack Sennett stuntwoman from Hollywood, the wife of a First Family of Virginia aristocrat, a Chicago stripper, a nurse-on-horseback from Kentucky, the niece of the Governor of Oklahoma and a member of the Florsheim shoe family who arrived with seventeen trunks and an escort of afghans and took over a suite of rooms at a Houston hotel. But as they headed for Houston, whatever their expectations for a women's Army flight training program, they were met on their arrival with a rather different reality.

Houston Municipal Airport, in 1942, was a vast tract. At one end was an enormous air base, Ellington Field, where thousands of combat pilots were being trained in almost everything from 450-horsepower basic trainers to B–17 Flying Fortresses. At another end of the airport was Houston Municipal itself, its hangars, tower and commercial airlines terminal. At the edge of one of the runways was a five-room shed and hangar belonging to Aviation Enterprises, Ltd., the new home of the AAF Women's Flying Training Detachment.

Actually, Cochran's WFTD would not have immediate access to the shed, only to the hangar. Aviation Enterprises had yet to complete its previous contract with the CPT program. Until the CPT cadets left, the Army's women pilot trainees would be without classrooms or eating facilities or even bathrooms at their flying school. For toilets and a cafeteria, they would make the half-mile trek to the Houston municipal airline terminal.

The flightline of the WFTD was the most motley assortment of airplanes in the Army—twenty-two different types of Taylorcraft and Piper Cubs, Aeronca trainers and other small airplanes, no two of which were alike. Most college CPT programs had bigger and better aircraft. The trainees greeted them as if they were rediscovering familiar childhood toys. The instructors, however, who were charged with training the WFTD to fly "the Army Way" and who had fewer hours in the air than some of the trainees, huddled apprehensively at one side of the hangar whispering among themselves, "Have you ever flown one of those things?"

Their first few weeks, the trainees would look as motley as their flightline. Though they were explicitly not promised uniforms, the flying clothes they brought from home were soon worn out by the rigors of daily flying. And what they put on in the morning they wore all day, as there was nowhere to change at Aviation Enterprises. Some 43–1s and later 43–2s went to department stores, bought men's trousers (pants for women were rare) and had them altered to fit their female frames. A couple of especially devil-may-care trainees bought Western blue jeans. The only regulation item at the WFTD was a hairnet, required by the director of flying to be worn at all times on the flightline.

Not until mid-December would a pile of GI flying suits finally arrive for the nearly ninety trainees at Houston. The Woofteds rejoiced—until they tried them on. The khaki zippered GI jumpsuits came, of course, only in men's sizes, and most of them were sized 40 to 46. The shoulders were so outsized and the pant legs so voluminous that trainees immediately called them "zoot suits."

Though Cochran had fought for and won control of her training program from the Air Transport Command, the Flying Training Command, within which she had to work, was a different matter.

The Gulf Coast Flying Training Center grabbed the first available captain, and ordered him to Houston on November 16, where he joined the arriving women pilot trainees in the Aviation Enterprises hangar. A few days later, Jacqueline Cochran flew to Houston to confer with her new base commanding officer. She could not be on base at all times, she explained. Mrs. Leoti Deaton was assigned there as staff executive. When Cochran asked what duties he wished performed by Mrs. Deaton, the captain answered, "Keep the girls happy and out of my hair."

Nevertheless, Jacqueline Cochran was adamant that the women pilots program be as disciplined and regulated as any male cadet program. Though the curriculum designed by Cochran and the CPT program officials was ready to be put into action, there had been no time to formulate rules and regulations for the WFTD. Cochran and Deaton took a copy of the Gulf Coast Flying Training Center handbook and drew up a quasi-military rulebook of their own. But the WFTD adjutant, presumably to gain favor among the trainees, but also demonstrating his disdain for the program's attempts at legitimacy, repeatedly assured the women pilots that as civilians they were under no obligation to follow any of the WFTD regulations.

Base command, if that be the term, misunderstood the women pilots' reasons for being in Houston in the first place. The WFTDs wanted very much to belong to an important wartime organization. Discipline at the WFTD was less necessarily by threat of courts-martial than by influence. As Mrs. Deaton would try to persuade them, to maintain a schedule such as theirs, physical fitness, sleep and moderation were essential to avoid collapse, if not ultimately to maintain ground school and flying grades so they would not wash out of the program. And as chaotic as their program was, trainees wanted to be at Howard Hughes Field more than anywhere else in the world. All that was needed to compensate for the hard work necessary in the military was to feel legitimately part of the heel-clicking, uniform-flashing pageant of wartime America.

But there was more to a military flight school than just learning to fly the Army Way, and even though the base command was unconcerned, someone else appeared who would whip the WFTD pridefully into shape. Neighboring Aviation Enterprises was an Army supply subdepot managed by Lieutenant Alfred Fleishman. Day after day, Lieutenant Fleishman watched the women trainees racing back and forth from the airport terminal to the flightline in a frantic mob. When he found out what the program was, he paid the WFTD a visit to see if there was anything he could do to help. Immediately he saw that there was a great deal. The WFTD was trying desperately to be military, but no one was telling them how. Soon young Lieutenant Fleishman was the WFTD's self-appointed drill instructor and for an hour each day, physical training or calisthenics director, as well as an invaluable staff assistant to Mrs. Deaton—all totally unofficially. Lieutenant Fleishman became a legend, immortalized in trainees' songs sung to and from the airport and soon to be part of the WFTD and WASP tradition:

Mine eyes have seen the glory of my biceps bulging out,
Mine ears have heard the story of Lieutenant Fleishman's shout
My teeth have felt the gritty sand that we all gripe about
The 319th flies on!

The young lieutenant understood more about Cochran's WFTD than just the value of physical fitness to a rigorous pilot training schedule. He understood the import of the fledgling women pilots program, and was among the first to assure classes of incoming trainees that their leaders in Fort Worth and at AAF Headquarters in Washington thought they were special. "You are part of an experiment which will do more to advance the cause of equality for women than anything that has been done so far," he told a roomful of freshly arrived trainees in February 1943. "You are going to fly Army ships as part of the first group of women engaged in a war effort doing this man's job, in the history of America." Having passed the AAF Form 64 exam, they were as much like men combat pilots as Cochran could make them; physically they could take tough pilot training. But they were going to have to become tough emotionally with the same alacrity as Fleishman would shape up their bodies. "This is no boarding school," he warned them. He also gave them a piece of advice. "There is a simple directive which we can give you about this and about Army life in general. If you follow it, you will find yourself in a much better physical and mental state. It runs as follows: 'If the Army can dish it out, I can take it.' If you adopt that attitude and go through school with it, you will become bigger and better women." But the trainees had to be tough not only for their own sakes, but for the good of the experiment, he told them. "When the first real bad break comes and it should develop that women can't take it, it might affect the whole program . . . You will have to stick out your chin and show them." For the group before him, this was all they needed to hear.

In March 1943, a new commanding officer, Major W. W. Farmer, was appointed to head the WFTD. A West Point graduate in his mid-thirties, Major Farmer felt that a command was to be commanded, not ignored, and suddenly Cochran's women pilots training program had a fourteen-hour-a-day, seven-day-a-week commanding officer. As evidence of his concern, and astuteness, he had Lieutenant Fleishman officially transferred from the Air Service Command to the Flying Training Command and assigned formally to the WFTD.

By December, the CPT cadets were gone from Aviation Enterprises and the WFTD no longer felt like transient intruders. Their trainee salaries provided them with $150 a month. The contractor opened a small mess hall and charged a dollar a day for all three meals, but the trainees wondered if the food, which was trucked over from Ellington Field and reheated, was worth the money. As each new class arrived at Houston, there were always several who pushed their plates away with ostentatious disgust. After two or three days of ground school, flying and Lieutenant Fleishman's calisthenics, however, they dove into their dinners like athletes at a training meal.

Also in December, the Woofteds moved into their tourist courts. Though they bore such names as Oleander Court, they were not garden spots. Some of the trainees discovered they were assigned two to a room that had only one double bed. So while the WAFs in Wilmington were buying their own uniforms, the WFTDs in Houston were buying their own beds. The humid Gulf Coast motels of Texas bred trouble for some of the more affluent women pilots. One evening, Kay Menges, who had left a lovely apartment in Tenafly, New Jersey, where she taught in a special high school for gifted students, returned to her motel room to find that water bugs had eaten all the satin flowers off her bathrobe. But Kay and the other WFTD trainees did not spend much time in their makeshift barracks.

Shortly after dawn, the trainees, who at the school's busiest numbered as many as 200, were picked up at their tourist courts by Army semitrailer trucks, which resembled and were soon dubbed "cattle trucks," and were driven the eleven miles out to Howard Hughes Field. There they lined up for roll call and marched, as Lieutenant Fleishman's finest, to the mess hall for breakfast. After breakfast, they marched to the flightline for the morning's flying lessons.

Though new Army trainers were beginning to replace the assorted civilian planes hastily assembled so that Cochran's training school could begin, Houston was not an ideal location for a flying school, especially during the winter months when the WFTD trained there. The fog rolled in off the Gulf, often without warning. Trainee Eileen

Roach, who entered the WFTD in February, was accustomed to avoiding the thick mists of Oregon, where she had gassed and scrubbed airplanes to earn money to fly. But she found the Gulf fog exasperating. Just as she got in the air to practice some solo maneuvers, the fog drifted over the field. Suddenly Houston Municipal was hidden completely from her view and she was in the air all alone above an impenetrable, cottony layer of clouds. After circling frantically and craning her neck over the side of her open cockpit trainer, she finally spotted a hole in the cloud layer. She pulled back her throttle and dove through the hole, just as the fog closed over above. Once she was safely on the ground and taxiing toward the hangar, the tower operator began to yell in her earphones, "The 'Field Closed' beacon has been on for fifteen minutes! Didn't you see it?"

The humidity was not only in the skies, but also under foot. Once they were organized enough to march everywhere they went, their shoes became coated with another half-inch layer of mud with each footfall. Being in Texas for the first time, some of the trainees bought cowboy boots. They were sturdy, weatherproof, comfortable, and to buy them did not require scarce shoe ration stamps. One morning, however, mechanics reported that six of the airplanes had heel-sized punctures in the wing fabric and cowboy boots were banned on the flightline.

At noontime, they marched from their airplanes to the mess hall for lunch, then marched to ground school held in a nearby building commandeered by Aviation Enterprises. All afternoon, they studied mathematics, aerodynamics, engine operations and maintenance, and meteorology—a complete Army cadet course with emphasis on cross-country navigation necessary for future ferry pilots.

During the afternoon, trainees often spent an hour sitting in Link trainers, which simulated instrument flying. Then they shook out their arms and legs and ran out onto the grass for an hour of Lieutenant Fleishman's calisthenics before dinner. After dinner, they finally had some time to study and rest. The cattle trucks drove them back into Houston at eight, and at ten, lights usually were out.

If the Tyrolean Orchestra's edelweiss- and awning-decorated bus did not give them away, telephone calls from the women pilot trainees soon let the city of Houston know that a very special experiment was under way at the airport. Though the trainees' life was spartan, the city was soon doing its best to make them feel at home. Local banks offered them unlimited checking accounts. The city's restaurants accepted last-minute reservations. Nieman-Marcus offered charge accounts, and its busy Antoine's beauty salon patriotically fit in a woman Army pilot trainee any time.

One afternoon, in January, a group of trainees caught a glimpse of what their future lives as ferry pilots held for them, after they finished their many long weeks of training at Cochran's WFTD. As part of their flight instruction, they were taken to the Houston Municipal tower to observe the tower operators. As the group descended from the tower and walked through the almost empty terminal lobby, they spotted a small figure lying prone on one of the terminal waiting-room benches. When they got closer, they saw a young woman wearing a flying suit. Her head lay on a parachute and she was fast asleep. "A WAF!" one of the trainees whispered.

"That's going to be you one of these days," chuckled their instructor, who had accompanied them to the tower.

The WFTD trainees reduced their stride and tip-toed reverently past the sleeping woman pilot.

One Afternoon, in late March 1943, the traffic pattern at Houston Municipal Airport resembled a buzzsaw as an assortment of Army trainers circled and dropped for takeoff and landing practice; the traffic pattern was also practice in avoiding midair collisions. The trainees had been told, "Keep your head on a swivel," a hundred times.

Suddenly, over the radio, came a commanding woman's voice. "This is Jacqueline Cochran. Clear the air, I'm coming in for a landing." The trainer planes appeared almost to stop in midair. One 43–2 trainee, Carole Fillmore, from Oakland, California, was on her final approach to the runway when she heard Cochran's announcement. She landed and taxied toward the hangar. Off to the north, Carole watched a stagger-wing Beechcraft speeding toward the field. Some of the students had pealed off to climb out of her flightpath, others kept circling, until Cochran could come straight in for a landing on the now empty airfield.

"Now, that's a grand entrance!" thought Carole Fillmore, as she watched the big plane land, its large propeller missing the concrete by inches as it whirled, and taxied toward her. With a vain attempt to wipe the grime from her face, she approached Cochran's airplane for her first face-to-face meeting with the famous woman aviator, her commanding officer. She had met Amelia Earhart in 1937; Cochran, she would learn presently, was very different.

The director of the Women's Flying Training Detachment

climbed out of the cockpit and down onto the runway. "Here, honey, you carry that," she said to Carole as they began walking toward WFTD's five-room headquarters, and handed her a full-length mink coat. At first Carole was annoyed, but as she held the soft fur she had to admit that it was pleasant to feel something nice for a change after weeks of ten-hour days in a zoot suit.

Jacqueline Cochran, not always in minks, visited her Houston training school periodically from her Fort Worth Flying Training Command office, to ask trainees what they liked and disliked about their program. But this trip had a specific purpose. She would make the announcement that the WFTD was moving. The women pilots training program, step-child of Houston Municipal Airport, was taking over an entire air base of its own, 375 miles to the northwest, in Sweetwater, Texas. Its new air base was called Avenger Field.

Four classes had entered the WFTD in Houston, and in all, 230 women would receive all or part of their training there. On April 28, five and one half months after their bewildering arrival the previous November, the class of 43–1 would graduate at Howard Hughes Field, Houston. Though original plans of October 1942 called for 500 women pilots, in January, the ferrying division of the Air Transport Command had requested 750 women ferry pilots before the end of 1943, and 1,000 during 1944. The Houston facility simply would not be large enough to hold her expanded program; classes numbering 125 women pilots had already begun training in Sweetwater. And Houston trainees were to move there, bringing their airplanes with them.

On Monday, April 5, 1943, members of the class of 43–4, partway through their second, or basic, phase of instruction, flew their first long cross-country trip like real ferry pilots, to deliver twenty-six Army BT–13 450-horsepower basic trainers, and themselves, to Sweetwater, Texas. Across the Texas sky, they formed two roaring square formations behind two BTs flown by instructors as flight leaders. Sixty miles south of Sweetwater, the formations landed at an air base in San Angelo, Texas, to stop for lunch. San Angelo long remembered the day when two dozen exhilarated young women pilots, wearing rumpled oversized flying suits and hairnets, sat in one corner of the air base cafeteria and whistled at every attractive GI who walked in the door.

X

Cochran's Convent: The Only All-Female Air Base in History

We are Yankee Doodle Pilots
Yankee Doodle do or die
Real live nieces of our Uncle Sam
Born with a yearning to fly
Keep in step to all our classes
March to Flight Line with our pals
Yankee Doodle came to Texas
Just to fly the "PT's"
We are those Yankee Doodle Gals!

CHARLOTTE MITCHELL stepped off the train at the Texas and Pacific station, Sweetwater, Texas, into a downpour. As the train pulled out, she peered through the hissing steam for other young women, but the glistening platform was empty. Within seconds her orange coat was soaked. California's sunny San Joachin Valley, her home, seemed like the other end of the world.

"Goin' to Avenger?" A man in a cowboy hat yelled from the open door of an old station wagon. Charlotte Mitchell, six feet tall, was obviously one of the women pilots who were gathering at the air base eleven miles out of town. Charlotte nodded and the man got out of the station wagon and took her suitcases. "Don't rain like this very often," he drawled. "Mud gets up to your ankles."

After driving along a desolate stretch of straight road, they passed a guardpost and pulled up in front of a long, one-story building. Charlotte ran sloshing through the red mud to the door. The driver, following, set her suitcases inside. "Good luck, ma'am," he said, looking at her strangely from under the brim of his hat.

* * *

AVENGER FIELD, Sweetwater, Texas, was 70 miles north of San Angelo, 40 miles west of Abilene, 70 miles east-northeast of Big Spring and 2,350 feet above sea level. Typical of the terrain of west Texas, the local vegetation comprised mainly mesquite trees, buffalo grass and a few greasewood. A northerly wind blew at a constant twenty-five miles an hour, often whipping up a dust storm that approached the field like a solid red wall. The thermometer in Sweetwater reached 100 degrees in April and stayed there for five months.

Avenger Field looked like any of a hundred Texas air bases. Long, low, wooden buildings painted gray, clustered at the edge of two criss-crossing, 3,000-foot gravel runways. From the air, the boundaries of the field seemed to be drawn by the narrow taxiways connecting the ends of the runways with one another and with the hangars and flightline. Avenger Field was shaped like the state of Texas.

Of the air base buildings, eight were lined up one after another like railroad ties, running north-south. These were the trainees' barracks, which were divided into two six-woman bunk units called bays.

Another longer building with windows all across the front ran parallel to one of the runways. This was the "ready room," where the trainees waited for their turn to fly; it opened onto the flightline of airplanes. At one end of the building were several classrooms. During the day, half the trainees were in the ready room or flying; half were in class.

On the edge of the runway, a control tower was being constructed. At the far end of the row of barracks, a large hangar was also under construction.

Atop the administration building, where the new class of trainees was gathering, would eventually be a ten-foot wooden disc, rising like the sun from its slanted roof, and painted with a giant rendition of "Fifinella," the mascot of the AAF woman pilot trainee. Designed in the fall of 1942 by Walt Disney, Fifinella was a shapely female gremlin, wearing a short red dress and red boots, gold tights and helmet and blue goggles. Her two bright blue wings were spread behind her, and she was poised as if just about to touch down on the administration building's roof. But this was March 1943, and not for several months would Fifinella adorn the first building seen upon passing through the entrance gate. Charlotte Mitchell and her classmates of 43–W–5, 127 strong, were the first full class of women pilots to receive start-to-finish training at the 318th AAF Flying

Training Detachment at Avenger Field, the only all-female Army air base in America's history.

As the class of 43–W–5 stumbled dazed and aching through its first week at Avenger Field, the U.S. Second Corps and the British Eighth Army were plodding across the sands of Tunisia in an attempt to push Rommel and his Nazi forces back through the Kasserine Pass and out of Africa. In Russia, the siege of Stalingrad had been broken a month before, and the Russian army was advancing south and west to recapture Kursk, Kharkov and Vyasma. In the Pacific, the U.S. Navy was destroying a Japanese convoy in the Battle of the Bismarck Sea off New Guinea. And in Sweetwater, Texas, 127 women were finally flying for their country.

CHARLOTTE MITCHELL opened the door of the administration building to find a large room. Huddles of young women, maybe fifty in all, were sitting on the floor and drying out in clouds of cigarette smoke. As she started to walk around the room, she heard bits of conversation. "Where have you been flying?" "How many hours do you have?" "This is really something special." She began to approach one group but their eyes looked apprehensive, too much like hers. A few feet away was another group talking loudly and laughing; she decided to join them. They had all met on a train that had collected women pilots in Boston, New York, Pittsburgh, Columbus, Indianapolis, St. Louis, Kansas City, Oklahoma City and on through Texas. By the time it reached Sweetwater, they had taken over an entire car. Soon the East Coast contingent was teasing Charlotte about her California orange coat and she, more relaxed, teased them about their Eastern-drab tweed suits.

Each travel-weary, rain-soaked trainee who straggled in was eyed with interest. Then came the eager questions, "Where are you from?" "What were you doing before?" Answers ranged widely: a home economics teacher and manager of the ski team at Cornell; a mother of three from Pennsylvania; a secretary at Cessna Aircraft in Wichita, Kansas; a journalist from The New York *Times*; a wife from California whose husband had been missing in action since Bataan; a copilot on the national airline of Peru.

Suddenly the door opened and a striking young woman with dark hair entered and looked around the room from under eyebrows drawn in two high arches. Charlotte thought she looked familiar. In several minutes, she drifted over to Charlotte's group. She was a model from California, she explained in a soft, throaty voice. She might look familiar because she had posed as the Dragon Lady for cartoonist Milton Caniff, who drew "Terry and the Pirates" and "Steve Canyon." Charlotte was awed. Like all the women pilots in the room, she was an ardent reader of Caniff's comic strips about flying.

The big room, known as the rec hall, had filled up to over a hundred young women. The high-pitched buzz of voices quieted down as a handsome Army officer, in his mid-thirties, appeared and began to speak.

He was Colonel Landon E. McConnell, commanding officer of Avenger Field. "Welcome to the class of 43–W–5 of the 319th Army Air Forces Flying Training Detachment," he announced. The young women beamed. Then his manner turned stern. "Look on each side of you," he said. "Only one of the three of you will graduate." One hundred faces fell. If they had come to Avenger Field for glamour, Colonel McConnell told the roomful of soggy young women, they had better take the next train home. They had joined the Army, and for the next twenty-four weeks, they would be told when to do what, how to do it and when they could stop doing it. Their living quarters would be subject to military inspection. A dime had to bounce off their bedsheets. Demerits would be given for infringements, starting tomorrow. He did not want to hear any profane language, Colonel McConnell told his women cadets. Finally, he told them that for the first two weeks they would be confined to base, as preventative quarantine. Then he wished them good luck.

The room was somewhat subdued at the end of this speech, but then Leoti Deaton began to give them instructions in her warm Texas drawl, and they perked up. They were told to choose five others with whom they would share a bay, then to line up outside the laundry window. Charlotte Mitchell and the California model for the Dragon Lady, Nedra Harrison, joined the East Coast train contingent. They lined up and were tossed sheets, pillowcases and blankets from out of the linens window. Then they ran through the rain down a long walkway to take possession of the very last bay. They had all agreed they did not want to feel boxed in.

The six baymates stood in the middle of their new living quarters. The standard Army cots were so narrow that Charlotte wondered if she would have room to turn over. One of her new roommates, Virginia Acher, from Terre Haute, Indiana, who was as tall as she, wondered the same thing. "We're here to see how adaptable we are," said Virginia, who held a Masters degree in personnel administration and whose father, a psychology professor, had studied with Freud at Clark University.

At the foot of each cot was a double-sized footlocker, which would serve as closet, dressing table and bureau. At one end of the bay was a door leading to the bathroom, shared with the adjoining bay of six trainees. It contained two stalls, two sinks, two shower heads and a mirror. "For this they dock us $1.65," Charlotte joked.

Suddenly they heard barking outside the bay. Charlotte stuck her head out the window, which was a board propped open with a stick, and saw a young woman talking emphatically with a young man in khaki. In one hand she held the leash of a black poodle, in the other a lit cigarette in a long black holder. "I wonder if you could tell me *where* I'm to put my things and what I can do with my little *dog* here." The woman wore a mink coat, rather matted from the rain, and the biggest diamond bracelet Charlotte had ever seen. The young GI argued inaudibly and the two turned and walked back toward the administration building. Charlotte pulled her head back in the bay and laughed to herself. Then they all undressed somewhat self-consciously and climbed into bed. Charlotte lay still for several minutes listening to the sleepy breathing of five people whom, a few hours ago, she had never seen before, then exhausted from her cross-country trip, finally fell asleep.

WHEN REVEILLE sounded shrilly at 6:15 the next morning, the diamonds, dogs, tweed suits and orange coats were gone, along with most of the trappings of individuality. The trainees were issued parachutes, which strapped uncomfortably between their legs. They also got flight caps, goggles and GI flying overalls—the now infamous "zoot suits"—which they wore into the showers and then hung on the clothesline out in back of the bay, hoping in vain that they would shrink.* One wasp-waisted trainee, trying to eke out the last drop of sex appeal, cinched in her zoot suit belt buckle so tightly that, climbing into the cockpit of a PT–19, she cracked two ribs.

Before the sun set, at 1830 hours by the Army's twenty-four-hour clock, the class of 43–5 would be an instant military unit. The trainees were divided into two flights, and each elected a flight lieutenant in charge of calling her flight to breakfast formation and marching them in columns everywhere they went for the rest of the day. One, Mary Parker, had been a private school teacher from Oxford, New

* Not until the spring of 1944 would WASP trainees be issued regulation WASP flying clothes, including flight suits that fit (with dropseats).

York; the other, Helen Dettweiler, from Washington, D.C., had been the 1938 women's Western Open golf champion.

Every minute before the 6:45 breakfast formation was spent making their beds as tightly as they could. They found that a dropped dime actually *would* bounce. Then they cleaned the bay and fought for mirror space as they pulled pin curlers out of their hair and put on make-up. Cosmetic attempts did not last long. Within two weeks, when Mary Parker shouted, "Fall in! Two minutes!" outside the bay, they had just awakened, and were splashing water on their faces, tying bandanas around their heads, and zipping up their zoot suits. They had mastered all the short cuts to snatch an extra half-hour of sleep.

For meals in the mess hall, the trainees lined up cafeteria style. Spooning from steamy service troughs was a smiling, gray-haired woman who wore wire-rimmed glasses and was called "Mom." On the night of their first air base dinner, the women were ravenous. They were delighted when Mom dished out steak, string beans, mashed potatoes, rolls and butter, and, for dessert, ice cream. The trays were demarked by different sized indentations, and by the time they reached their tables, the ice cream had melted over into the potatoes. But Army rations were generous. Most of the new arrivals had not eaten real butter (which was rationed for civilians) for many months.

After dinner on their first full day at Avenger Field, the weather cleared just in time for the sky to turn from rose to orange to crimson. Charlotte and her baymates lined up outside the bay and watched. In spite of the rain, mud and desolation, they realized they were in Texas because of its wide, spectacular sky.

An hour later, writing "I don't know where to begin . . ." letters home as they lay on their cots, they heard a shout. "Attention!" Into the bay stormed a group of trainees from 43–W–4, the class ahead of them. In mid-February, half of 43–4 had started training in Houston, half in Sweetwater; the latter seventy-five trainees were about to assert themselves as the pioneer veterans of Avenger Field. Charlotte, Virginia Acher and the others hopped off their beds and stood up straight. "This is the Army, wipe that smile off your face!" they were ordered raucously. Like the new recruits on Army air bases across America, members of 43–5 were about to be hazed.

Trying to wipe the smiles off their faces, they were made to march up and down in front of the bays. Then some of them were thrown, fully clothed, into the shower. Suddenly Charlotte saw a hand heading toward the cold water tap, and shouted, "Everyone out of the way!" in time to escape the scalding spray.

Finally, the initiates lined up panting, laughing and dripping water across the middle of the bay. Then they were freed, as their tormentors moved onto the next bay.

The next morning, the base commanding officer made a formal announcement that henceforth all hazing of new trainees at Avenger Field would cease. One of the women trainees had been admitted to the base infirmary the night before with a splintered tailbone, suffered when the chair she was ordered to sit on was pulled out from under her.

SINCE 1942, Avenger Field had trained pilots from the war-torn United Kingdom, and when 43–5 poured into Sweetwater, the last class of UK cadets had yet to graduate. For two weeks, the men and women flew at opposite ends of Avenger Field and their regimented schedules kept them apart. But their living bays were next door, across a no-man's land only a few feet wide. The women trainees could hear men singing and playing ball. At night, hamburgers cajoled out of Mom were mysteriously passed across no-man's land from the cadets to their quarantined neighbors. Then the cadets were gone.

In the first week of its existence as an all-female air base, over a hundred curious male pilots from the myriad neighboring cadet training schools in west Texas made "forced landings" at Avenger Field. Then the base command ordered the field closed to all air traffic except in an emergency, and Avenger Field became known as Cochran's Convent.

Soon the tables were turned. One night, a fierce electrical storm caught three cadets on a cross-country flight. They circled Avenger Field for half an hour before they could land. When they could finally see the runways, they dropped down to safety. The pilots soon forgot their fright. As they climbed exhausted from their airplanes, they were surrounded by dozens of grinning young women, who, upon hearing engines, had hopped out of bed, dressed in their zoot suits, and rushed in an eager mutiny to the flightline. Fifteen minutes later, Mrs. Deaton appeared and sternly shooed the women back to their bays. The fun was over. Such an audacious display would be most damaging to the image of the women pilots program.

The "Army Way" of submerging sexuality, the trainees soon

found, besides the rumored dash of saltpeter in the stew, was to keep them physically *tired*. After four hours on the flightline and five hours of ground school, they would be called out to the playing fields for a rousing game of baseball or volleyball. "I'll be so slim (athletics and such), tanned (the Texas sun) and neat (the Army)," one trainee wrote home to Oregon. They were soon in better shape than ever, and no one was there to see.

But 43–W–5 had come to Avenger Field to fly. Though the trainees all had pilots licenses, their flying experience had usually been limited to 40- and 60-horsepower Cubs. Compared to Nancy Love's WAFs, many of Cochran's women, though they had demonstrated their enthusiasm for flying, were real beginners in the air. By their second day at Avenger Field, they were climbing in and out of the 175-horsepower Fairchild PT–19s as if their primary phase trainers were Rolls-Royces.

Soon they were taking off, three at a time, with their instructors, in the little open-cockpit trainers. As more classes arrived at Avenger Field, there would be as many as fifty planes in the air at one time. As the tower had not yet been completed to regulate air traffic, 43–5 trainees developed the eyes of hawks in a matter of days. In the traffic pattern, they flew stacked four or five levels high as they circled and dropped down to the next level until it was their turn to land.

From the air, they saw that west Texas was perfectly laid out for flying. All the roads were navigational aids—they ran north-south and east-west. As trainees approached the Avenger Field runway, however, their troubles began. One of the runways was almost always under construction, so from their first day of training every landing was cross-wind, or perpendicular to the wind. They had to learn to dodge tiny cyclones filled with dust, called dust devils, which danced around the runways and could tip up a wing in a second. Also hazardous on landing were meandering flocks of tumbleweed which could jump up at an airplane to become entangled in the landing gear, or entwined in the propeller. The trainees did not know it, but they were getting the best of flight training, mainly because of the habitat.

They also learned that their daily physical training classes were more than just for discipline or for sublimation of sexual energy. One of the calisthenics involved rolling their heads dizzily around on their necks, first one way and then the other. The trainees thought the exercise was silly. Then they were made to do their first spin in a PT–19. When the wing stalled and dropped through the air, the plane whipped them around so fast that they felt as if their heads would fly off. Fortunately, whiplashes were avoided—their neck muscles had been well prepared. After two weeks' instruction in the PT–19, they began one by one to solo. Soloing the PT was the first milestone of training. Every day as they marched to the flightline, singing from a growing repertoire of Women's Flying Training Detachment songs, they passed a large round reflecting pool in the middle of the base filled with three feet of water. The pool was called the Wishing Well. In the bottom shimmered coins thrown in for luck by trainees about to take primary check rides. When a trainee first soloed, she was tossed, zoot suit and all, into the Wishing Well.

After soloing, trainees were allowed to take the PTs up by themselves to practice areas surrounding Avenger Field. As the weather became more summery, the open cockpit of a PT at 3,000 feet was a pleasant place to be.

One day a trainee could not resist taking a sunbath. She soon learned that, though Cochran's Convent was off limits to male cadets, in the skies trainees were fair game. She trimmed the PT perfectly so that it flew itself, held by the stick between her knees. Checking for other aircraft in the area and seeing none, she took her shirt off and leaned back, her face pointed toward the sun. After several luxurious minutes, she heard an unexpected roar. Opening her eyes, she saw two other PTs on either side of her. Soon she was surrounded by a flock of primary trainers. In the cockpits were not women trainees from Avenger Field, but male cadets who were grinning and waving enthusiastically. Her PT began to weave and bob as she fumbled with her shirt in the windy cockpit. Then the shirt slipped out of her hands and sailed out over the Texas plains. Above the roar of the engines, her aerial audience cheered. She ducked down in the cockpit, and sneaking looks over the side, banked steeply and headed back toward Avenger Field. When she landed, she taxied to the flightline, cut the engine and from a seemingly empty cockpit, waiting trainees heard a voice yell, "Somebody bring me a blanket!"

The instructors of Avenger Field, who were civilian employees of the War Training Service and attached to the AAF, were not always objective when faced with a female trainee. Resentful and disconcerted that Avenger Field had switched from male to female cadets, some instructors decided to treat their young women students like men. When a female trainee did something wrong, they yelled. Trainee Virginia Streeter, a twenty-eight-year-old former corporate nutritionist in New York, vowed she was tough enough to take it.

She hopped into the front seat of a PT, her instructor climbed into the back seat. She put on a helmet and earphones through which she could hear his instructions over the roar of the engine. She, however, had no microphone. Communication between instructor and student was obviously supposed to be one-way.

They flew out of the Avenger traffic pattern and high over the Texas plains. First Virginia was told to do a power-on stall, then a power-off stall, and finally some coordinated turns. Throughout the stalls, he yelled that she was not pulling the nose up high enough or that she was losing too much altitude before recovering a safe flying speed; on the turns she was using too much or too little rudder. By her second turn, he was yelling "How can you be so dumb!" so loudly that she pulled her earphones half off her ears. Then she realized that the insides of both knees were beginning to hurt. With each mistake, the stick which was jointly controlled from the front and the back seat, would jump from her hand and knock hard against her knee. She could feel bruises forming.

After half an hour, she had had enough. She tore off her helmet and earphones, throttled back to quiet the engine, rose up and stuck her head back out the side of the cockpit toward the instructor. "Stop being so smart," she shouted. "I'm doing the best I can." From then on, the voice in her earphones, though not cordial, was businesslike. He never yelled at her again.

As an all-female cadet air base, Avenger Field had unique training problems. Personality clashes between Army flying students and instructors occurred on all flight training bases. The Army's attitude was "that's tough." The cadet put up or washed out. At Avenger Field, the student-instructor relationship was more complicated. The most vehemently reiterated rule at Avenger Field was that students were not to socialize with instructors. Casual chats on the flightline were inevitable, but no dates were allowed. (Several trainees married their instructors after graduation, which proved that rules were made to be broken; but their trysts had to be discreet.) Sharing the cockpit for many intense hours always created its own special authority-approval relationship, and emotions were consistently high as pressure mounted toward check rides during each phase of training.

For the first three months, trainees could not look to the base command for an objective hearing, either. While trainees who socialized were under the threat of discipline, base command felt free to maneuver. At the end of May, Jacqueline Cochran finally asserted her authority in the AAF Flying Training Command and had the entire Army command removed. For two weeks, Leoti Deaton ran Avenger Field until the command of another Texas cadet base could be found. The switch was made on June 7, and Avenger was then headed by Major Robert K. Urban, a forty-one-year-old experienced

Army officer who had been an Army Air Corp pilot and flight instructor since 1928.

A trainee review board had been inaugurated for trainees who received unsatisfactory flying grades from Army check pilots found to be overly stringent with female cadets (one check pilot came to be known as Captain Maytag because he washed out so many trainees). But in the switchover of base commands, one trainee was lost, and her plight dramatized the more subtle difficulties between check pilots and trainees. She was washed out in a check ride by her flight instructor before her official Army check ride. When she appealed to the review board, comprised of Mrs. Deaton, the new commanding officer, Major Urban, and a staff assistant (a former trainee who had washed out from Houston in February but was hired by the program administration), she claimed a personality conflict with her instructor. In truth, she never had been able to understand, much less comply with, his instructions. She was talking flying, but he wanted to teach her other things. The instructor throughout the hearing righteously defended his judgment about her flying ability. The review board knew that she had entered the program with among the highest number of hours of flying experience in the class and during proceedings became aware of the bitterness of a man rebuffed by the pretty trainee.

After the hearing, Major Urban ruled that she had to be washed out. Well advised of the ambiguity of the circumstances into which he had stepped, he nevertheless had no Army regulations to cover a case like this. He gratefully accepted a recommendation by Mrs. Deaton. If a trainee got a pink slip from her instructor, she would be allowed five extra hours of instruction with a different instructor. Though it was too late for that trainee, an uncomfortable fact of the woman's life of Avenger Field had been faced.*

WITH THE BEGINNING of primary training in the air, the female cadets also commenced ground school. Before graduation they would receive over 400 hours of what was the equivalent of a college aeronautics degree. They studied physics, aerodynamics, electronics and instruments, engine operations and maintenance, meteorology, navigation, military and civilian air regulations, and mathematics. When they sat down for their first class in this prodigious course, however, the

* The trainee was called to basic instructor's school and became the only woman in the AAF who taught basic students. WASPS were put through instructors training in 1944, but were not allowed actually to teach students.

ground school instructor gave them an arithmetic test, "Just to see if you girls can do any of this stuff." Though they all had passed the Army's most demanding intelligence tests, many men were still skeptical about whether women could master the technical exigencies of flying. At Avenger Field, where examinations were given two or three times a week, not one woman washed out of the pilot training program because she failed ground school. The female trainees breezed through mathematics; but, belying assumptions about women's clerical facility, they dreaded having to achieve the speed of six words a minute when transcribing each month's new International Code.

CHECK RIDE TIME in mid-May was charged with emotion for 43-5. They had been given check rides by their flight instructors, who were civilians, but to finish training, military check pilots flew into Avenger Field and took trainees up for official Army flight tests. Before each check ride, trainees ran to the Wishing Well, tossed in a coin, and said a little prayer for an "up-check" to Fifinella. Those who passed ran happily to a big fire bell by the administration building and rang it in triumph and relief.

If a check ride was unsatisfactory, a trainee was given a pink slip with the ominous letter "U" as a warning. After a couple of days' practice, she was then allowed to take a second check ride. If she failed that, she washed out. She had paid her way to Sweetwater and she would now pay her way back home again. Washing out, most of them felt, meant they had lost the biggest opportunity of their lives to do something of importance and to be part of something special.

The ready room on the flightline was usually a place where a lot of Coca-Cola, cigarettes and philosophy were consumed. But as primary training culminated in final check rides, the tension began to mount. Check ride time was like an air show, as the trainees were put through the complicated and beautiful aerial maneuvers that were designed to save their lives in an emergency. The difference was that the air show spectators at Avenger knew their names might be called next to perform.

One afternoon, Charlotte Mitchell stepped outside to look at the on-forever Texas sky to give herself some perspective and to watch the show. In the distance she noticed a tiny white speck. A few minutes later, a PT landed downwind, in the opposite direction from the other airplanes. Something was obviously wrong. The PT came skidding up to the flightline and a check pilot vaulted out of the open cockpit. His eyes were wild. "Lost my girl!" he shouted. "She fell out!" The back seat of the PT was empty, the gosports hung over the side. "She took us into a spin and suddenly she wasn't there any more!" Charlotte looked off toward where she had seen the white speck and now clearly discerned a parachute drifting down. A jeep roared off in the direction of the parachute and everyone paced the flightline nervously for half an hour. Finally the jeep returned, with the trainee and the parachute crumpled in the back.

The twenty-five knot wind had carried her four miles, and she had been dragged fifty feet, losing her shoes along the way, before she could spill the billowing chute. Her feet were bare, and her face scratched, but she wore an embarrassed grin and was waving her parachute handle. As she was surrounded by concerned trainees, she explained, "I popped the stick and before I knew it was soaring out over the windscreen. My seatbelt latch came unfastened."

Her instructor gathered her up, took her immediately to another PT–19 and made her fly it for half an hour so that she would not be spooked by the bizarre incident. In the mess hall afterward, however, she was hailed as Avenger Field's first bona fide member of the Caterpillar Club, the prestigious group of airmen—and now women—who had safely bailed out of an airplane and had their parachute ripcord handles as proof.

That afternoon, Charlotte heard her name called, and two hours later she had successfully graduated from the PT–19 to the BT–13 basic trainer. She returned to her bay to find one of her baymates in civilian clothes. "The check pilot told me I looked too frail to handle an airplane," she told Charlotte numbly. As tears smarted in her eyes, Charlotte sank to her cot and watched her baymate transfer clothes from the footlocker to her suitcases. She had become close to the dark-haired, soft-spoken woman, a fellow-Californian. The Dragon Lady had washed out.

The next morning, a cot sat glaringly vacant in Charlotte's bay. No one could talk about the tragedy that had occurred. But Charlotte and her remaining baymates made a pact. Like a comic book fighter squadron, they would always think of themselves as the Invincible Five.

Having to pass the rigorous Form 64 Army Air Forces physical exam before being accepted to Avenger Field, Jacqueline Cochran's women trainees compared favorably to the most highly selective combat flying squadrons of the war. But a few of the physicals had been less than thorough. Charlotte Mitchell, who was working as a telephone operator at the Ted Rankin Flying Academy in Bakersfield, took her AAF physical at Gardner Field. Like many of the young women who suddenly appeared at Army air bases to be examined by Army flight surgeons in 1942 and 1943, she was the first female to whom her doctor had given the Form 64. The young flight surgeon was far more nervous than she, and Charlotte did not get a very detailed physical examination. He checked her eyes, did some blood tests and pronounced her fit. On another base in Pennsylvania, an entire ward was closed off temporarily so that two hopeful women pilot applicants could walk freely from lab to lab in their short, open-back hospital gowns.

Once at Cochran's Convent, trainees were examined freely, as walking, flying female guinea pigs, by the Army Flight Surgeon of Avenger Field, Dr. Nels Monsrud. At Jacqueline Cochran's direction, he monitored their every move. One thing he found was that the arbitrary height-weight charts adapted for female trainees were overweight by four pounds.

Assuming that weight inferred strength, minimum weights had been set for each age and height, roughly corresponding to charts for male cadets. A twenty-six-year-old woman pilot who was five-foot-seven inches tall had to weigh at least 125 pounds or she would not be accepted. (Standard weight for that height was considered to be 139, and the maximum acceptable was 175.) One aspiring trainee from San Francisco spent a long lunch hour at Fisherman's Wharf eating a nine-course meal before being driven up to a Marin County air base by her mother (who would not let her go to an air base alone) for her physical. In the car was a quart of milk and a bunch of bananas to eat on the way. She still weighed in three pounds too light. Thinking fast, she told the flight surgeon that she had been on a diet. "Well, you're overdoing it," he told her indulgently, and passed her. Once she got to Avenger Field and was more sinewy than ever, she wore a bandana full of buckshot to all of her periodic weigh-ins.

But Dr. Monsrud's mission was more exploratory than merely determining realistic height-weight charts for women Army pilots. He was making a pioneering and definitive medical study to determine if women could take the strenuousness and constant concentration of such a demanding professional activity as daily flying. For it, he ordered trainees to report all the normal incidents and dysfunctions of their menstrual cycles. Embarrassing as this research was to the trainees, many of whom had never spoken with anyone about such intimate details, Dr. Monsrud collected data leading to results that were, for the era, revolutionary.

Comparing trainees' flying grades to their menstrual cycles, Dr. Monsrud compiled a unique document called "Medical Considerations of the Women's Airforce Service Pilots," which he submitted to the Air Surgeon General in 1945. As he related in this report, Dr. Monsrud found no correlation between menses and elimination from training, nor was there any with high or low flying grades. No demonstrable menstrual factors were involved in any of Avenger Field's flying accidents. In fact, of minor accidents such as ground loops, only 28 percent occurred within five days after the first day of menses; the rest divided evenly throughout the month. For two classes specially studied the following year, he found that flying grades were slightly higher around the first day of their periods than on the tenth day in between periods. He concluded that trainees were trying especially hard on those days. But flight surgeons on air bases across the country who followed the trainees after graduation corroborated Dr. Monsrud's findings. Women pilots on duty lost *less* time per month from operational flying, when grounding themselves voluntarily for physiological reasons, than men pilots.

THE TOWN of Sweetwater, Texas, did not know quite how to view the invasion of young women who came to Avenger Field to fly airplanes. Sweetwater residents had enjoyed having the male cadets at their nearby air base because their daughters could socialize with the young men. But when Avenger Field no longer provided class after class of potential husbands, the dusty little cattle community lost interest. After a considerable public relations effort by Mrs. Deaton and her staff assistants, who wangled invitations to speak on the local radio station, the town finally resumed a long-standing custom with Avenger Field.

Though only a few streets wide, the town of Sweetwater contained twelve churches. Trainees were invited to attend services and then go to parishioners' homes for Sunday dinner. But at times, Sweetwater's gesture at hospitality proved to be half-hearted. Several trainees found themselves relegated to eat their meals alone in the kitchen.

In April, Jacqueline Cochran flew into Sweetwater and called on the mayor. He was so charmed that he agreed to organize a day of special events so that the town and the trainees could get to know each other. After an afternoon of songs and skits by multitalented trainees from Avenger Field, the young women were invited to nearby ranches for impromptu rodeos and barbecues. Sweetwater was won over. No longer were they eyed as highly suspect women of the world. "They're civilians, but men of the flying fraternity think they're the greatest little sisters that a group of Army fighting airmen ever possessed," enthused the Sweetwater *Reporter*.

The goodwill of Sweetwater was bestowed just in time. In mid-May, the AAF Flying Training Command issued its first national press release on the Women's Flying Training Detachment at Avenger Field. Graduation ceremonies would be held for the second class, 43–W–2, at Avenger Field, on May 28, 1943. Having successfully completed flying training at Howard Hughes Field in Houston, class members would fly their fast, powerful advanced trainer aircraft—single-engine 650-horsepower AT–6s and twin-engine AT–17s—to Sweetwater. Their exodus from Houston marked the closing of that base and the consolidation of all training of women pilots at Avenger Field.

The graduates of 43–2 had no uniforms to wear for the ceremonies. But as the WFTD's first class had done for its graduation at Howard Hughes Field in April, each one went to the Ellington Field post exchange and bought tan general's pants, short-sleeved white shirts and boat-shaped general's caps. Wings, all-important to air force cadets at graduation, were more of a problem. After much debate, the ever solicitous Lieutenant Alfred Fleishman had gone to the PX and selected bombardiers "sweetheart" wings. With Mrs. Deaton, he designed a shield with a stylized ribbon above, on which the squadron and class could be engraved. Then they gave the wings and drawing to a local Houston jeweler who re-created the shields in metal and soldered them to the middle of the bombardiers sweetheart wings. The bill went to Jacqueline Cochran.

As trainees of 43–5 sat on the flightline in their dusty zoot suits on May 27, they spotted the formation of advanced trainers in the distance. They got to their feet to watch as one by one the AT–6s and AT–17s zoomed in to land and screeched to a halt in front of the hangar. Young women hopped out of those huge ships, shook their hair out jauntily in the wind and waved to the awe-struck trainees on the flightline. Press photographers clicked and newsreel cameras whirred.

Then Jacqueline Cochran flew in from Fort Worth. She emerged from her plane and, followed by her French-speaking maid, began to walk toward the operations building as some of the Houston graduates respectfully surrounded her. While greeting Cochran, one graduate lost all five painstakingly acquired bottles of liquor, which she had hidden behind the seat of her AT–17, to a vigilant Avenger Field mechanic. But in a few hours she would hardly miss them.

The graduates were driven in Avenger Field "cattle trucks" similar to those in Houston, into Sweetwater, where they would spend the night at Sweetwater's only hotel, the Blue Bonnet. Jacqueline Cochran was given two rooms next to one another, the Blue Bonnet's approximation of a suite.

While the 43–5 trainees studied and then slept at Avenger Field, the Blue Bonnet rocked with celebration. Around midnight, the last of 43–2 arrived from Houston, having driven almost 400 hot, dusty miles in a graduate's car. One of the tired travelers was Carole Fillmore. Out of a window of the hotel someone shouted her name in an inebriated greeting. When Carole checked in at the hotel desk, she was handed a note saying that Jacqueline Cochran wanted to see her. Carole had met Cochran only once, when the famous female record holder had flown into Houston and asked Carole to carry her mink coat. How she knew her name, or why she might want to see her that night, Carole could not fathom.

When she stumbled wearily into Cochran's room, she found out. The director of women pilots training immediately began to yell. "I've devoted my life to you girls these past few months and I have been so proud of you and now you all are ripping this place apart!" Cochran had just walked in to find one of the graduates in her bed. "How can you do this to me?" Carole was nonplussed. "It isn't my fault," she said. "I just got here. They just yelled my name out the window." She sat down on the bed. "I've been driving all day, I'm hot, and I'm tired. I haven't had a damn drink. And I could use one." This time it was Cochran's turn to stare at the young woman, a trainee and now graduate of her Women's Flying Training Detachment, as if seeing one for the first time. "I guess you deserve it," Cochran said finally. She got a glass from the bathroom, reached in a drawer and pulled out a bottle, and poured her exhausted graduate a tall drink.

The next day, 43–W–5 showed off their marching ability in the graduation parade review. With great exuberance the long columns, headed by a flag carrier and the two flight lieutenants, marched past Jacqueline Cochran, several high-ranking officers from the AAF Flying Training Command in Fort Worth, and the commanding officers of the two Women's Flying Training Detachments, the now historical 319th from Houston, and the 318th going strong at Avenger Field. Bringing up the rear was the Big Spring Bombardier School Band.

As they paraded past 43–2 graduates, Charlotte Mitchell had to bite her lip to keep a straight face. She noticed that some of the graduation celebrants from the night before were leaning precariously against one another and blinking painfully in the bright Texas sun. "Lucky devils," Charlotte thought admiringly. She had only just begun basic instruction, and her own graduation ceremony seemed very far away.

THE AIRPLANE flown during basic flight training was the BT–13. After the PT–19, the BT seemed enormous with its enclosed canopy top and its 450-horsepower engine. Though the manufacturer appellation was the "Valiant," basic flying students called it the "Vultee Vibrator." When put into a spin, it shuddered as if it would shake apart in midair. To pull it out of the spin, it took almost 2,000 feet of altitude before it would fly level again. Unlike the primary trainers, however, the BT–13 had a radio.

The Avenger Field tower was completed just in time for this training advancement. Those trainees who had not flown at major airports before the war, or had not worked as control tower operators, had to learn an entirely new language. "FF eighty-one from fifty-six on the ramp requesting taxi instructions . . . over. . . ." they said into the microphone now attached to their earphones. "Fifty-six in number one position, ready for takeoff . . . over. . . ." No longer did they rely merely on their swivel-heads, eyes and guts; they would be told what to do and when to do it by the control tower operator.

In basic instruction, they also learned to fly "under the hood." A skill rarely acquired even ten years before, when Jacqueline Cochran was learning to fly, "blind" or instrument flying was now part of every air corps cadet's curriculum. Because the 43–5s were training to be ferry pilots, they would have more instrument time than male cadets in training, who, presumably preparing for combat, received more acrobatics instruction. The women of Avenger Field no longer flew by the seat of their zoot suits. Instead, they flew for hours under a black curtain, with their instructor yelling in their ears, and with their eyes riveted to the airspeed indicator and altimeter to see if they were climbing or descending, or the needle-ball to tell them if they were slipping to the left or right. Instrument flying was grueling, concentrated work.

Although as ferry pilots they would fly only between sunup and sundown, part of their instrument training was to experience flying at night, should they run out of daylight on a mission and have to land after dark. As Avenger Field had no lights, burning oil pots were placed all along both sides of one runway to make the outlines of the landing strip visible.

Trainees going night flying had to make a mad rush for the showers before lights out, to wash off the day's Texas dust and to wake themselves up. Then until two in the morning, they would practice taking off and landing. These simple procedures were completely transformed without the perceptual contrasts revealed by daylight. Taking off was like hurtling oneself into a dark void, with no feeling of speed. Only their airspeed indicator told them when to break ground and, robbed of a horizon, they had to trust their indicator that they were climbing at a fast enough, safe speed. After hundreds of greased-on landings, suddenly they were falling to the gravel runway with a devastating thud. Spread out before them in the landing headlight of their BT, the runway just was not where it seemed to be, and the oil pots flashing past their eyes were mesmerizing. After a few touch-and-go landings, their instructors took them off into the dark Texas night to practice navigation by radio beam. For those on the flightline waiting around for their turn to fly, Mom would set out sandwiches and coffee.

Getting the feeling of night flying held many surprises for the basic phase trainees, who now boasted of almost seventy-five hours in the air since they arrived at Sweetwater. But one night they ran into one of Texas' natural phenomena. The entire field was engulfed in a cloud of crickets. The crickets covered the sandwiches and hopped into the coffee pots. If the canopy had been open too long between occupants, trainees would climb into a cockpit hopping with crickets. The next morning, when they got out to the flightline, they were given brooms. In half an hour they swept several two-foot mounds of dead crickets off the ramp.

Night flying was exciting and a difficult challenge, but it was to bring Avenger Field its darkest hour since the arrival of women cadets. On June 7, baymates Jennie Hrestu and Jane Champlin hopped into the shower before a long evening flying BT–13s out over the blackened Texas plains. Jennie had just heard from her family in Portland, Oregon. Her two little daughters were fine. When, in January 1943, Jennie's husband could not get out of his war-essential job, he called a family conference. The four Hrestus had recently emigrated from Greece. Many of their relatives had been killed or captured during Nazi aggressions. Someone, her husband said, should be sent to help fight fascism. Since Jennie knew how to fly, she was the one to go—to ferry airplanes that would be used in combat in

Europe. Jennie was delighted that she would get to serve her new country.

Jennie Hrestu's baymate, Jane Champlin, had come to Sweetwater from St. Louis, where she had been a secretary in the Railway Express Agency. As soon as she heard about Jacqueline Cochran's training school, however, she quit her job and flew almost full-time to get the required number of hours. Then she had invested all of her savings in two expensive nose operations to correct a sinus and breathing condition in order to pass the AAF physical.

As Jennie and Jane scrubbed life back into their weary bodies, they both decided that night flying was fun. They could picture themselves on night bombing raids with the RAF over Germany.

Out on the flightline, Jennie and Jane waited their turn to fly. Finally, waving jauntily to each other, they followed their instructors out to BT–13s and one after the other took off into the Texas night.

Jennie flew navigational problems and practiced landing and taking off for about two hours. Finally, about one in the morning, she came staggering back to the bay with one thought in mind—the oblivion of sleep. She looked over at Jane's cot. It was still empty. But Jane's instructor had the reputation of occasionally getting off course, and Jennie sighed for Jane. "He probably got lost," Jennie thought. Like several of the instructors at Avenger Field, he was only a couple of weeks ahead of the trainees in learning to fly the planes in which he was teaching. This instructor also had a tendency to turn bright red and to scream irrationally at students at the slightest error.

Thanking her lucky stars her instructor knew how to navigate, Jennie fell into an exhausted sleep. When she awoke early the next morning, however, Jane's cot was still empty. At breakfast formation, everyone was talking in low worried whispers.

As the morning group arrived at the flightline, operations told them that the bodies of Jane Champlin and her instructor had been found with the charred wreckage of their BT–13 on a ranch near Westbrook, Texas. The engine apparently had caught fire in the air, and they had been unsuccessful in extinguishing it before the BT–13 crashed.

Many of the trainees began to weep. Avenger Field's first fatal air crash immediately sank deep into their fears. Jennie, feeling numb, sat against the ready room wall. She had been the Hrestu sent to war; now the tragedy of war had hit. Jane Champlin could have been any one of them, in the wrong airplane at the wrong time. But most of them had not directly faced that fact of the hazardous duty they had chosen to perform for the war effort.

Suddenly, the voice of one of the instructors rose above the sobs that were filling the ready room. "All right, enough of this," he shouted. "There's a war on, girls, and we've got to get flying." The first flights wiped their eyes and followed their instructors out to the airplanes. They all knew that was the way it had to be.

After the fatal BT accident, the flightline changed tenor, as if the trainees, though still motley in their baggy zoot suits, pigtails and turbans, began to take their mission—and one another—more seriously. Soon after they soloed the BT–13, and had flown many hours under the hood with an instructor, an announcement was made on the flightline. They were to choose a "buddy" with whom to take up a BT and practice instrument flying. One would pull the black curtain hood over her seat and fly the airplane on instruments while the other, as the instructor had done, would watch for other aircraft and be ready to take control of the stick at the split-second of an emergency.

The choices were quickly made. Flight Lieutenant Mary Parker teamed up with Pat Pateman, a writer who had been working in a defense plant in New Jersey before joining Jacqueline Cochran's flying training program.

Talking casually about compasses and gyros, Pat and Mary walked out to a waiting BT–13. Suddenly they found their insides were fluttering as if they were on a first date with their dream man. They realized that while they had flown with an instructor and alone, in all of their training they had never flown with one another.

Mary and Pat took off. Once far from the field, one let the other practice banking, climbing and descending while she checked the nose altitude, in case some instrument was not working, and scanned the environs so they would not collide with anyone. Then they switched. Halfway through the flight, Pat realized that her life literally was in Mary's hands. Instrument flying was very concentrated work, and if Mary made a serious mistake, Pat would have to take over immediately. But to her surprise, she felt calm, even smug that she and her friend were up in their own airplane, far from their instructor.

Soon they were back on the ground. As Pat and Mary started walking toward the ready room, each spontaneously threw an arm around the other's waist and burst out laughing. "You know, Parkie," Pat said, "you're a helluva pilot!" "Pateman, so are you!" Mary said almost in unison. For two months, they had studied together, flown with the same instructor, been bored together in the ready room, discussed flying problems together and had become very good friends.

Suddenly they both realized something else about one another that they had never shared—they were not only great gals, but also crack pilots. By the end of the afternoon, the ready room had a new electricity as teams returned to the ground; the admiration and trust among them gained in those few hours would grow stronger, eventually encompassing all the women pilots who would be flying with the AAF.

THE LAST STRETCH of advanced training specialized in experience they would need as ferry pilots—long-distance navigation. The airplane was the North American Aviation AT–6 Texan which, with its pug nose hiding a 600-horsepower engine and its A-shaped mullioned canopy, remained many pilots' favorite airplane. The AT–6 had retractable landing gear, making it appear very sleek on takeoff, and the decreased drag pushed its cruising airspeed to 145 miles an hour.

In the AT–6, 43–5 got its first full taste of the woman Army pilot's world beyond the gates of Cochran's Convent. Their long solo cross-country flights for navigation practice taught the future ferry pilots how important a sufficient gas supply was—too many sputtered to a midair standstill and then glided shame-facedly into the nearest cotton field or pasture, miles from any ranch house.

They also learned that their reception at air bases away from home was not always cordial. By July, they knew every little cow town in west Texas from the air, especially along their cross-country practice "triangle" of Odessa, Big Spring and Sweetwater. One day Virginia Streeter, freshly dried from being thrown into the Wishing Well after soloing the AT–6, decided she would vary her day's cross-country trip by reversing the triangular course. Her instructor shrugged in approval but warned her to pay attention.

Two hours after taking off, she realized nothing below looked familiar. Not knowing where she was on her maps, she had no frequency to dial on her radio, so she followed the old standby "iron beam." Shortly she came to a depot and dipped down to read the sign. It read, "Frederick." Five miles away, she spotted a big airfield, and without calling the tower—Frederick, Texas, on her map showed no air base nearby—she came straight in for a landing. The AT–6 was still a hot ship to her and she landed twenty to thirty miles an hour faster than she should have. Nevertheless, her fast, power-on landing looked like that of the most experienced fighter pilot.

She taxied up to operations. "Where am I?" she asked cheerily as she walked in. "Frederick, Oklahoma," the officer growled. "Came in awfully hot, didn't ya?" Virginia realized she was a hundred miles off course, and completely off her maps.

"I'm lost," she said.

"I figured as much," replied the officer glaring at her. "We had one of your kind here last week." He picked up the telephone and called Avenger Field. "We've got one of 'em up here," he said over the line. Then he turned to Virginia and said, "You can go."

Virginia walked back out to her AT–6 feeling chilled to the core and apprehensive about the homeward trip. He had given her no maps, no headings, no directions of any kind, and she was afraid to ask. She took off and followed some railroad tracks heading southwest and just as night was falling, she reached Sweetwater. Her instructor met her on the flightline. Virginia humbly admitted her error of overconfidence. She also had gained new respect for the established cross-country triangle out of Cochran's Convent.

But several weeks later, on a much longer cross-country, Virginia met with accolades, not annoyance. The trainees at Avenger Field, unlike male cadets, were not always the object of pride within their families. After one trainee from Maryland joined the WFTDs, she heard nothing from her mother for eight months. Another, at home in St. Louis on leave, watched her aunt cross to the other side of a downtown street to avoid having to speak with her.

Virginia had paid her way to Sweetwater from New York City, where she had worked for AT&T as a dietician. She had been at Avenger Field for three weeks before writing to her parents in Cedar Falls, Iowa, about what she had done. Her father had denied her permission to fly airplanes; flying was not a proper profession for his daughter, or any girl. If she insisted on joining a service, he told her, she might join the lady Marines. Now *there* was an elite corps.

Halfway through advanced cross-country training, Virginia landed in Des Moines to refuel. While her AT–6 was being serviced, she debated whether she should fly up to Cedar Falls, as a reconciliation gesture to her parents. Finally she picked up the telephone.

Forty-five minutes later, she entered the traffic pattern at Waterloo airport near Cedar Falls. She looked below; her grandfather had settled all that land before the Civil War. A small airport crowd gathered at the fence was gawking at the hot Army advanced trainer zooming around for a landing. A man in the crowd shouted, "Look what's coming in!" The AT–6 landed fast, wheels greasing onto the runway, and taxied up to the gate as the crowd emitted ahs and oohs of admiration. Mr. Streeter finally could not hold in his pride. "That's my daughter," he said to his neighbors at the fence, and strode over to the plane. The canopy slid open and the five-foot-three-inch young woman with long curly brown hair rose smiling from the cockpit. Mrs. Streeter looked over her daughter's shoulder. "Where are the rest of the people in there?" she asked apprehensively. But Virginia had flown home to Iowa by herself. As her father extended his hand to help her off the wing of the AT–6, Virginia realized she had banished disapproval, if not incomprehension, forever.

THOUGH THE CLASS of 43–5 was far too busy to reflect on its import, a new development occurred in the organization of women pilots throughout the Army Air Forces. On July 5, 1943, as 43–5s were soloing AT–6s in Sweetwater, the War Department in Washington issued a press release. Women pilots in the AAF, those in training and those flying for the ATC ferrying division, were now under the sole jurisdiction of a Special Assistant and Director of Women Pilots assigned to the Assistant Chief of Air Staff, Office of Commitments and Recruitments. The title would be known as, simply, Director of Women Pilots, and the War Department press release announced the appointment of Jacqueline Cochran to the job. WAFs head Nancy Love, the release continued, would direct the women of the Air Transport Command.

There were now 120 WAFs and WFTD graduates on active duty with the Army Air Forces. By September, after the graduation of 43–5, there would be well over 300. General Arnold, at Jacqueline Cochran's urgings, had determined that there must be a person at AAF headquarters to oversee women pilots' activities, especially since they were bound to embark on important flying duties other than ferrying, as Cochran and he had long intended.

A month after Cochran left Fort Worth for her new office in the Pentagon, the AAF's women pilots were renamed. No longer WAFS or WFTD, now they were all to be called Women's Airforce Service Pilots—or WASPs.

AS ADVANCED TRAINING check rides approached, Sweetwater, Texas, became unbearably hot. The temperature had been hovering around 100 degrees for weeks, and the trainees were working and flying with singed nerves. One trainee was put in the infirmary for a breakdown, caused by the perpetual hot wind. Many, in spite of their fatigue, could not sleep. As the bays ran in the same direction as the prevailing wind, there was no cross ventilation. When sleeping in wet towels did not help, trainees were given special permission to drag their cots out into the walkways between the bays.

Every once in a while giddiness would set in. One day, between the bays, some of the women rigged a hose in the slats of a straight-back chair. Putting down boards on the dirt, which was soon mud, and donning bathing suits, they took turns running through the nozzle spray like children.

As graduation approached, almost everyone caught a dose of "convent fever." Each day a colored flag was raised on the flagpole to indicate what dress was appropriate. A green flag meant that flying gear was the order of the day; a red flag indicated that civilian clothes were allowed, as on a Sunday. One day, the flag pole bore both. Manic for entertainment, the trainees appeared in formation wearing hats with veils, high heels, fox furs and dressy coats over their zoot suits. Another morning, rather than playing "Reveille," the bugler searched her repertoire and woke the reluctant trainees with the old Army cavalry call, "To Horse!"

On days off, pressure was released by a visit to a room above the local giftshop in Sweetwater where, for $2.00, a gray-haired woman would reach under a checkered tablecloth into a basket, and pull out a bottle of clear liquid. (Sweetwater was in a "dry" county.) Then they could go to the new Avengerette Club, contributed by the town to the AAF's little sisters, and drink spiked Coca-Cola until the ten o'clock curfew. One warm summer night a trainee made it back to base in time, but not to her bay. In the morning, she was found sound asleep in a ditch behind the building.

To add to their physical discomfort were the hours they had to spend in the Link trainer. Flying was joyous relief because the temperature dropped 3.5 degrees with each 1,000 feet they climbed. Flying with ventilators open was like stepping into an air conditioner. The Link trainer was called the sweatbox. It looked like a toy airplane ride at an amusement park, but it was not so much fun. It had stubby wings absurdly affixed to the outside. Inside was a cockpit, with earphones connected to the instructor's table in the front of the room, and instruments electronically wired to simulate flight. This was necessary training, but after an hour in the sweatbox, the pilots were all too ready to climb out into a "cool" 100 degree wind.

At last, graduation was scheduled for Saturday, September 11. At the end of August, with two weeks to go, the last phase of advanced training began—checking out in the twin-engine AT–17. The airplane did not have a very good reputation, and the 43–5s joked callously about the plywood plane with its boxy fuselage. Though its manufacturer, Cessna Aircraft, called it the "Bobcat," fliers called it the "Bamboo Bomber," or more ominously, the "Bunson Burner." The AT–17 looked as if it could go up in flames in seconds. The trainees thought they were hot enough pilots to talk about airplanes in that manner, never dreaming an airplane's reputed idiosyncratic tendencies might make any difference to them. What they did find peculiar to the AT–17 was that when the instructor took up two trainees the one who sat in the back seat, sunken behind the pilot and copilot, often needed to have a paper bag handy.

One afternoon, a week before graduation, several trainees and their instructors took AT–17s on cross-country flights to practice navigating by radio beam from Sweetwater to Big Spring, seventy miles away. Graduation was so close that everyone was feeling exuberant already, and navigation practice took on new import as they knew that soon they would be ferrying AAF trainers by themselves as AAF Air Transport Command pilots. Late in the afternoon, the AT–17s began to arrive back at Avenger Field for supper. As the trainees and instructors strolled talkatively into flight operations, they were met with the news that Helen Jo Severson, Peggy Seip and their instructor were dead. Their AT–17 had crashed near Big Spring. The flightline was in a state of shock. By dinner, word was circulating that their Bunson Burner had blown apart in midair. Just as they were about to get their wings as AAF pilots, the women of 43–5 had become considerably more superstitious about using pilots' jargon nomenclature for the airplanes they flew. (An official investigation for the accident report later revealed that structural failure may have caused the crash.)

After dinner, Charlotte Mitchell and other trainees gathered around the rear wall of the neighboring barracks where Helen Jo and Peggy had lived. A leafy vine the two had planted in March now covered the wall completely. Helen Jo Severson and Peggy Seip, of the Class of 43–5, had created their own memorial garden. As Charlotte looked sadly at the thriving vine, she vowed to water it and keep it alive until graduation day.

Zoot suits and parachutes
And wings of silver, too,
He'll ferry planes like
His Mama used to do!

* * *

SATURDAY, September 11, 1943, eighty-five members of the class of 43–5 marched in stately columns past the long line of AT–6s toward the reviewing stand at Avenger Field. Standing respectfully at attention as they passed were the classes of 43–6, 43–7, 43–8 and the newest arrivals, barely into primary training, 44–1, for whom Avenger Field would be home until February 1944. West Texas had given 43–5 its finest weather for graduation. The scorching sun had mellowed, the temperature had dropped into the eighties, and the wind had a fresh smell of impending autumn.

One by one, as their names were called, they filed up to the reviewing stand to meet Jacqueline Cochran, who smiled as she pinned on their silver WASP wings. For Charlotte Mitchell, as for them all, this was one of the proudest moments of her life. The 190 days she had spent at Avenger Field seemed like years measured in experience, friendships, sorrow and skill she had gained as a WASP trainee. But she also felt she had proven something more than just a personal accomplishment. She felt as if she were part of a triumphant experiment. Two thirds of her class had made it through those many weeks of rigorous flight training, as many or more than at an Army male cadet school. The silver wings over their left breasts proved women could fly Army airplanes. They would go on to prove that women could do a "man's job" ferrying for the Army Air Forces.

As Charlotte stood at attention while the Big Spring Bombardier School Band played the Air Corps song, she had no inkling that Jacqueline Cochran had broadened her experiment far beyond an air base in west Texas. She and several of her 43–5 classmates were about to be assigned to forge the next frontier.

XI

When Women Flew the Flying Fortress

On October 15, 1943, Lieutenant Logue Mitchell left the dinner table in his Columbus, Ohio, apartment to take a telephone call. The twenty-four-year-old B–17 flight instructor at Lockbourne Army Air Base was annoyed. He had just graduated his tenth class of pilots in the four-engine bomber and he was looking forward to a rare quiet evening at home with his wife, Lois.

"You have some new students showing up tomorrow, I want you on the flightline at seven sharp," said Major Fred Wilson, Lockbourne's squadron commander. "By the way, six of them are girls."

"I've heard rumbles. Why me?" asked Lieutenant Mitchell.

"Well, I'm not sure I can keep sex out of the picture, but I guess since you're happily married there will be as little as possible." Major Wilson's voice rippled with humor. He was a thirty-year-old bachelor. "There will be seventeen women in all. We'll let Major Hurley worry about the rest of them. He has two kids."

Ever since Logue Mitchell was selected to check out in the early experimental YB–17 version of the Flying Fortress in 1942, the young flight instructor was always given the tough students—the cocky, the fearful, the near washouts. He knew that women were flying for the Army. He remembered hearing their voices over the flight radio when he was stationed at Ellington Field, Houston, in the winter of 1942. Now he was going to have to teach them how to fly the famous Fortress.

Back at the dinner table, he told his wife about his new students. "You've always said 'You don't have to be a superman to fly a B–17,' " she said. Her husband grinned. "It'll do the guys good for them to find that out," he said.

After being stationed for only two weeks with ferrying groups at

New Castle, Delaware; Romulus, Michigan; and Dallas, Texas, thirteen members of the class of 43–5, and four from 43–6, who had just graduated from Avenger Field, received orders to report for special assignment to Lockbourne Army Air Base in Columbus. All of them were baffled. In the train to Columbus from Detroit, WASP Virginia Acher was asked by several men pilots where she was going. When she answered "Lockbourne," their mouths dropped open. "What are you going to do there? It's a four-engine school."

"Is it?" Virginia asked. "I guess we're going to sit in the tail and spit out the window to test wind direction."

When Helen Dettweiler, Dawn Rochow, Mary Parker and Charlotte Mitchell arrived at Lockbourne from New Castle, they were told to report immediately to the flight operations building. As they approached, they gaped at the flightline. It was a long row of the biggest airplanes they had ever seen, mammoth four-engine bombers 75 feet long, with wingspans of over 100 feet. The three-bladed propellers were almost twelve feet in diameter. "What are we going to fly?" asked Frances Green. Way at the end of the flightline was a tiny L4 Cub. "That must be it," said Charlottle Mitchell.

In the operations office, six of the women pilots were greeted by a gangly young officer, over six feet tall, with an affable face and large, warm blue eyes. "I'm Lieutenant Mitchell, your flight instructor," he said. "Let's go out and meet the 'Big Friend.' "

"Then it's true!" shouted Mary Parker as they climbed around and through the huge airplane. Lieutenant Mitchell told them cursorily what everything was, but as they gazed at the instrument panel, packed with gauges, dials and switches, they all became silent with awe.

"We're going to get every one of you through this," Logue Mitchell said.

"You're going to have to work your tails off," said Frances ("Greenie") Green, grinning. Startled, Lieutenant Mitchell looked at the dazzling pretty blonde who had uttered the comment in a Texas drawl.

The other eleven WASPs were gathered across the room, around Major John J. Hurley, Lockbourne's deputy director of flying training, and Major Larry Berglund, an instrument instructor, who would teach them how to fly the Fortress. "Here we are, Hurley and Students, Inc.," said Dawn Rochow, nervously. All seventeen WASPs and their instructors were going to work hard. Flight training in the Fortress took 130 hours in the air and the WASPs would get the same training as the male officers at Lockbourne, of which there were over 300 going through the B–17 AAF transition school. First they all would learn how to operate a four-engine airplane, how to synchronize the engines, how a plane that big stalled, and other basic flight maneuvers. Then they had to learn how to fly and land the Fortress when one of the 1,325-horsepower engines, or even two, went out. Meanwhile they would have many hours of cross-country experience, by day and night, to get intensive navigation and radio experience. Though the Fortress could hold ten hours' worth of fuel, no operations office wanted to entrust a $300,000 weapon of war to a pilot who did not know where he, or she, was every minute. They would be on the flightline five hours a day, and they would often be in the Fortress for four of them.

Three days earlier they had been learning Air Transport Command procedures to deliver PT–19s. Now, their first afternoon at Lockbourne, each was taking turns sitting in the first pilot's seat of a 50,000-pound B–17 Flying Fortress bomber. Lieutenant Mitchell realized that each time he spoke to one of his six new students she became rigid with fright. "You all have the same problem," he told them at the end of their orientation to the B–17. "Putting up with me until you've learned to fly this airplane."

After dinner at the officers' mess, the WASPs unpacked in their rooms, provided them for $6.00 a month, at Center Hall, the nurses' quarters on the base. Suddenly a bottle of wine and some glasses were produced. They toasted to their amazing good fortune. "Why us?" they asked. Then they took a hard look at one another. They averaged five-foot-eight inches tall, and two of them were six-footers. "I guess it's because we're the larger type," said Charlotte Mitchell sipping her wine. In fact, Charlotte was right. Cochran had ordered that the seventeen WASPs assigned to B–17 school be five foot four, and preferably more.

"I feel like Alice in Wonderland," said Charlotte. They soon began to feel quite giddy as the reality hit them that they were about to be pilots of the biggest, most famous bomber of the war.

The next morning, they rose at five. And at seven-thirty Mary Parker, a twenty-five-year-old former physical education major at Russell Sage College, was adjusting her live-rubber earphones onto her ears in the pilot's seat of her first Flying Fortress. With Lieutenant Mitchell next to her, she went through the check list, as the other WASPs gathered around behind the pilot and copilot seats to listen. The Fortress was big enough to hold all of them for each flight, though they sometimes went up three or four at a time.

After turning on the electrical switch that activated a hydraulic pump to give oil pressure to start the number one engine, far to the left, they came to the item that read "Start Engines." Mary pushed the button and the huge eleven-foot seven-inch propeller shuddered and began to spin, powered by a 1,325-horsepower engine. In a minute and a half, all four engines were thundering.

Then, hand on the engine throttles, she followed Lieutenant Mitchell's movements by feel as the gigantic bomber began to move forward. After only two weeks flying the "Bamboo Bomber" AT–17, with its two 225-horsepower engines, the Flying Fortress felt like taxiing a hotel. Maneuvering the two outer engines, they wove forward to the end of the runway and with a roar of one outboard engine, Lieutenant Mitchell swiveled the bomber into position, heading straight down the runway. "This thing'll never get off the ground," Mary said dubiously. "That's what they all say," said Lieutenant Mitchell. Following his instructions, she edged the four throttles forward in her fist and the bomber began to roll. Mary watched the airspeed indicator as it climbed. At 110 miles per hour, she pulled back on the control wheel, and the huge bomber lifted smoothly off the runway. The other WASPs, strapped into seats behind her, were mesmerized. As they reached the end of the runway, she pushed the control wheel forward, bringing down the plexiglass nose so the B–17 would gain speed. Then, with a hum, the main landing gear and tail wheel came up into the belly of the plane and when the airspeed reached 135 they started to climb. As they eased away from the runway, they looked out the windscreen at the wings that seemed to extend forever to each side. Lieutenant Mitchell closed the cowl flaps, which had helped cool the engine on takeoff. Now the Fortress was roaring ahead. Mary pulled her earphones off one ear. The noise of the engines was deafening and she had never felt such a sensation of power in her life.

After takeoff, Charlotte and Greenie stood up in the ball turret behind Mary to watch and listen. Charlotte could not get over being able to stand up straight in a cockpit. But as she looked around her, anxiety again returned. It would soon be her turn to take off in the behemoth "Big Friend."

Once they left the traffic pattern, they flew out over the Ohio countryside to practice some banking turns. Every maneuver seemed almost ludicrously monstrous, and the WASPs laughed nervously at the exaggerated size of the wings and nose jutting out of the cockpit. But Lieutenant Mitchell insisted the Fortress handled just like the small airplanes they were used to flying. He climbed to 12,000 feet and dazzled them by putting the Fortress through a perfectly coordinated slow roll. Then he put the bomber into a spin. "Just like a dancing elephant!" shouted Mary, feeling lightheaded.

When she got back to her room that evening, Charlotte would write home to reassure her family in Tulare, California, that their children were all right flying an airplane of such inconceivable size. "Tell Aunt Bertha," she wrote, "she needn't worry about Hugh in a B–17. They're really the safest ships built, but just big."

In the following days, the WASP pilots found that re-entering the traffic pattern at Lockbourne was the most unnerving maneuver. Already circling the field, like a school of whales, going 140 miles per hour, were several other B–17s. On their first few landings, the women pilots bounded down the runway like bucking broncos. A three-point landing, when the wheels were fifty feet apart, seemed impossible. The Fortress hit on its main gear, then bounced back on the tail wheel. As the nose went up, the pilot pushed the control wheel forward and the Fortress hit the main gear again and bounced higher back onto the tail wheel. Finally the bomber slowed down enough to settle onto all three wheels.

Learning to land an airplane weighing between twenty and twenty-five tons (depending on the armaments it was carrying) was hazardous not only to the airplane but to the WASPs' precarious reputations. Helen Dettweiler once dropped a ship so hard onto the runway that she broke a wheel strut. As she walked past mechanics on the ground, she was berated mercilessly. There were 125 Fortresses on the field and the engineering department had a contest going to see how many they could keep in service each month. She had just ruined their record, they claimed. They knew the advent of girl pilots at Lockbourne was bad news.

The mechanics were not the only field personnel who were distrustful of the new female students. On each training flight, a flight engineer would ride along to monitor the B–17's electrical systems and engine performance. For the first couple of weeks, when Major Fred Wilson issued assignments for the day, the flight engineers would grumble, "Do I hafta?"

Lieutenant Mitchell was aware that his WASPs were not particularly welcomed by the other pilots who had come to Lockbourne to train on the B–17. "What's this war coming to," he heard one say to another, "that they're giving girls the same training that we're getting?" The officers going through four-engine school at the 1174th Squadron at Lockbourne were an impressive group of majors, captains and lieutenants, many of whom had already seen combat. As the seventeen newly arrived WASPs walked into their first ground school class, they passed a rack strewn with officers' hats. Then they sat down, shoulder to shoulder, with the men. Along one wall of the small square white room was a display board mounted with airplane parts.

For the first few minutes there was a feeling of tension in the room as the officers eyed their female classmates, both with appreciation and uneasiness. After the instructor welcomed the WASPs to Lockbourne, Helen Dettweiler, who had been the Women's Western Open Golf Champion in 1938, rose from her chair. "Colonel, we are here to do our best," she said, "but we want this understood. We want to be treated like the other students here. If we pass, all right. If we don't too bad." This assertion cleared the air. For three months, from one to five o'clock every afternoon, the class of seventeen WASPs and their officer classmates would study every academic aspect of the Flying Fortress.

As women in the classroom were no longer disconcerting, the WASPs began to be accepted; so much so, that Dawn Rochow wrote home to her parents in Rochester, New York, that she was ambivalent about the new relationships she was forming with her fellow B–17 students. Being taken seriously was a new experience. But when her officer friends began to call her a "good fellow" she was not sure she liked it. One of her dates told her he got a big bang out of talking to her: she could understand what a manifold was, and how a carburetor worked. She knew if the fuse for the tail wheel was in the main fuse box or in the secondary fuse box. "Always the pal!" she wrote wistfully. "And yet, I've decided that is the best way to be." Most of the officers with whom the WASPs were training would be sent to Britain to fly Fortresses for the Eighth Air Force over Nazi-occupied Europe, and forming strong romantic links with them seemed like playing with fate.

In four-engine school in 1943, both the men and women confronted the same bewilderingly complicated aircraft. During their engines course, they had to tear down one of the B–17's 1,325-horsepower Wright engines. The ground school class divided up into groups. Greenie, two other WASPs and a young lieutenant found themselves facing a huge Wright R–1820–27 nine-cylinder radial air-cooled engine. Dismantling the myriad parts and laying them out on the table was fun. But when they had to reassemble the engine the three WASPs, assuming looks of helplessness, asked the lieutenant repeatedly what to do next. Finally, sweating in exasperation, he shouted, "I don't know either!" When he suggested a valve lifter and inserted it upside down, they all mustered their courage. The engine had to run or they would all fail. When they got the engine put back together, Greenie spotted two ball bearings lying on the table behind the assembled engine. She discreetly slipped them off the table and dropped them in her pocket as the engine coughed to a start and roared triumphantly. She kept them with her as lucky pieces.

When the WASPs and their fellow male students completed the engine course by mid-November, the women earned among the highest marks. But although they had done well on their exams, Virginia Acher and Charlotte Mitchell re-enrolled when the course was repeated the following week to increase their confidence. Though they continued with the chart-reading phase on the regular ground school program, they did not mind taking on the extra work. That was what they were there for, to *know* the Fortress.

By mid-November, Lieutenant Mitchell and his training flight of six WASPs had established a routine. He waited in the ready room, drinking coffee and keeping warm, and sent the WASPs out to walk around the Flying Fortress. He watched them, looking like vigilant penguins in their fleece-lined winter flying suits, as they checked the tires for air pressure, the turbines for hydraulic leaks, the wheel struts and break clearances for wear. They pulled themselves into the metal fuselage, icy cold from a long night on the ramp, and checked inside to make sure the emergency crank was fastened in the bomb bay (in case of gear failure) and that nothing was loose around the radio room. Then two of them sat in the pilot and copilot seats to check the gauges on the instrument panel. When Lieutenant Mitchell saw them through the cockpit windows, he put down his coffee cup and ran quickly through the chill morning to the plane. As soon as he climbed through the fuselage to the cockpit, he nudged the WASP from the copilot's seat and nestled into the warmed cushion.

Lieutenant Mitchell was determined that the Fortress was going to be as familiar to the WASPs as the main street of their home towns. On every flight, he gave them "blindfold cockpit checks." With their eyes closed, they had to be able to point instantly to every button, knob and instrument. To learn cross-checks, they also had to fly with his hand covering up any of the instruments. They learned to tell by the sound of the engine whether they were ascending or descending.

He taught them how to take off only by instruments. Greenie was first. She soon learned that a "hooded takeoff" was as sinister as it sounded. Around the windscreen Lieutenant Mitchell snapped a black tentlike curtain, completely covering her view. He instructed Charlotte to stand behind them once airborne and watch for air traffic to the left. He taxied the plane into takeoff position on the runway, and locked the tail wheel straight ahead so the rear end of the plane would not fishtail on the takeoff run. "She's yours," he said. Then he stuck his head under the curtain and yelled, "RELAX!"

Greenie took the throttles in her fist and gently walked them forward. She felt the plane begin to jostle and bounce under her. "I guess I'm moving," she said. Then she riveted her eyes to the airspeed indicator and watched the needle begin to move down the circle—40, 50, 60 miles an hour. Suddenly she realized she had to watch the directional gyro compass and make sure she was still on the runway and not aiming the bomber off onto the grass. The plane rumbled and bumped along and finally was at 110. She pulled back on the control wheel and the bouncing stopped; the Fortress felt smooth, almost as if sitting still, except that the airspeed indicator was at 135 miles per hour and the foot-needle of the altimeter was rotating. As soon as she reached 1,000 feet, Lieutenant Mitchell was unsnapping the curtain and barking about watching the gyro. Like magic, the city of Columbus appeared below.

From their second day in the air, Lieutenant Mitchell had them reconnoiter with other Fortresses to learn the skill, and thrill, of flying in a formation of huge roaring four-engine bombers. "Get in closer!" he would shout. All the WASP at the wheel could imagine was bumping the wing of the plane next to her and having part of it fall off. Her first instinct was to peel off and fly in the opposite direction. "See how you're bumping around in all the prop wash?" he would shout. "Don't look at the horizon, don't look at your instruments, fly on that plane!" They soon found that the closer in they got, the smoother the Fortress flew, as they edged into the same air currents as the plane beside them. Flying in formation at night was even more spectacular. Instead of the wing, the pilot fixed on a little blue flame emitted from the exhaust pipe of the next plane, keeping the point of light in the same spot on the windscreen.

Since Lieutenant Mitchell believed that the WASPs did not have the brute strength of his male students, he worked hardest on perfection of flying technique. The more precisely the pilot flew the Big Friend, the fewer the strains. He began by teaching them to hold the throttles from underneath, palm up; there was more strength in their biceps, elbows and wrists, he explained, than an over-the-top grip, which demanded strong shoulders. He also made sure that they were getting good physical training on the ground. They were taught how to take pages from the newspaper at one corner and crumple them into their fists. They vaulted horizontal bars. They squeezed Bernard McFadden wrist developers. In their rooms, they developed their arm and shoulder muscles by lying under the cots and pushing them up off the floor.

The flight maneuvers demanding the most strength were two- and three-engine procedures—holding the Fortress level with one or both engines out on one side. Lieutenant Mitchell had watched 200-pound officers step from the airplane and crumple to their knees with exhaustion after a morning of two- and three-engine procedures, Lieutenant Mitchell told his WASP students, but he wanted them to be able to come down safely should they lose an engine. The only way to learn was to practice.

A WASP would take off, get to 1000 feet, make her first turn, and Lieutenant Mitchell would cut an engine. Out on the wing, a propeller would be windmilling in the airstream. She would stomp on the rudder pedal to correct the imbalance in thrust, adjust the other three throttles and get the plane straight and level. Then he would cut out the second engine on the same side, and he would make her fly all the way around the traffic pattern on only two engines. After twenty minutes, she would have sweat pouring off her face; her rudder pedal leg would be trembling uncontrollably with exertion. Then he would call back in the plane for the next WASP victim. Each one underwent this particular trauma many times.

One day, after going through several rounds of two- and three-engine procedures, Lieutenant Mitchell and his drained women pilots landed and taxied up to the ramp. He heard them all breathe sighs of relief; they were finally going home for a hot shower. They were wrong. "Okay, parachute drill!" Lieutenant Mitchell yelled. The WASPs stumbled to their feet, grabbed at their chutes, slithered them over their shoulders, and, bumping into each other as they tried to get through the cockpit door, scrambled back through the fuselage to the hatch. Lieutenant Mitchell strolled back to the panting line-up and checked his watch. "Not bad," he said. "One minute forty-five seconds. There is only one thing wrong." The WASPs' faces fell and their eyes followed as his long finger pointed at one of the chutes. Every one of them had put on her chute upside down. Lieutenant Mitchell climbed down out of the fuselage and, looking back through the hatch, dismissed them with a downward wave of his hand. His prize WASP students were rolling, dissolved in giggles, on the fuselage floor.

But as torturous as two- and three-engine procedures were, Lieutenant Mitchell had made sure his WASPs could do them well. He taught each of them a trick. When the leg depressing the rudder pedal was shaking so hard it was about to lose control, she could hook the toes of the other foot in back of the unused rudder pedal and pull up. The leverage relieved the pressure just long enough to rest the first leg, so it could hold a few minutes longer. This trick might save her life, he exhorted. (A few months later, when several WASPs were stationed at a Florida air base, the trick would save their pride.)

Though released from Cochran's Convent, the WASP B–17 pilots were still watched over with great interest by the Army medical corps. First, an anthropologist came to Lockbourne to measure their skulls. Then, halfway through their flight training, they were taken to Wright-Patterson Field in nearby Dayton to be tested in a high-altitude chamber. It was standard procedure to test B–17 pilots for their reactions at high altitudes, as well as to prepare them for actual in-flight experience. Other than flight nurses who were beginning to work in hospital planes, the WASPs were the first females to fly at 25,000 feet and above in unpressurized cabins. The AAF was concerned about the effects of high altitudes on women.

For twenty-four hours before their trip, they ate (supposedly) only cottage cheese and drank milk. Lieutenant Mitchell and six WASPs joined several of the other WASPs and Major Hurley and they were all flown to Dayton. In the flight test center, they were ushered into a small room the size of a cockpit with benches along each side and a window at one end. They were instructed to remove watches from their wrists and pens from their pockets.

The first test was to demonstrate the effects of oxygen deprivation. They were given pads and pencils and told to write "Mary had a little lamb," or "My name is . . ." as many times as they could. Then the chamber door was bolted shut. They started writing. Suddenly men were holding oxygen masks over their faces, the pencils were drooping and their hands were lying on their knees. The last line of writing on their pads fell lazily, illegibly off the paper. None of them had felt anything but a cheerful sensation before she went out.

Then the chamber was sealed again, and the pressure was decreased to simulate flying at 30,000 feet. One of the WASPs soon found that she had forgot to remove her fountain pen; there was a pop, and ink splattered all over her shirt. Helen Dettweiler, who, straying from her bland diet, had eaten a breakfast of bacon and eggs that morning, looked down to see her stomach expand like a balloon. "I'm not really pregnant!" she protested as she caught everyone eyeing her inflating abdomen. Then, as the pressure climbed back to normal, Helen's stomach deflated. Finally the chamber door was unlatched and the WASPs walked out into civilization. To see how long the test had taken, one WASP glanced at her watch. All that remained was a tiny metal tray attached by a band to her wrist. It had exploded in the simulated high altitude, but everyone had been so busy with the other effects she was feeling that no one had noticed.

Though they lost a pen and watch, the WASPs gained cards which certified them as qualified by the Army to fly above 35,000 feet. They were also given cards to send back to Wright-Patterson over the next three years to report every menstrual period, instance of dismenorhea or amenorhea, pregnancy or miscarriage.

After their visit to Wright-Patterson, Lieutenant Mitchell took his WASP students up to 25,000 in the B–17. They all bundled up, tested the oxygen system serving the cockpit, took off and climbed for what seemed like hours. Once high up in the rarefied atmosphere, Lieutenant Mitchell told them to try some stalls. The rudder and ailerons felt as if they were moving in jello.

The world was an amazing sight from five miles high. Each took a turn lying on her stomach in the panoramic plexiglass nose of the Fortress. But Lieutenant Mitchell gave them a test to show that conditions were very different at that altitude in an unpressurized B–17. Beyond the closed cabin door was a narrow walkway above the bomb bays. On either side under the walkway lay two huge, inactive bombs. Beyond the bomb bays was another door which opened onto a narrow passageway through which one had to crawl to reach the tail gunner's area. Lieutenant Mitchell had each WASP take off her oxygen mask. Holding her breath, she had to leave the cockpit, run over the bomb bays, crawl through to the gunnery area and then come back, while Lieutenant Mitchell timed her. Then they had to make the same trip breathing oxygen from a carry-around bottle. The trip without oxygen, even though they were holding their breath and presumably rushing to get back to the cabin, took several seconds longer than on oxygen. "You can't trust yourself," Lieutenant Mitchell warned. They quickly gained respect for properly functioning oxygen equipment.

As their training progressed, the WASPs realized that they were practically living in a Flying Fortress. Surely it was big enough. But from five in the morning, they experienced and talked about little else. Their quarters at Center Hall, compared to the bays at Avenger Field in Sweetwater, were elegant. Each had her own room. Some had put pillows on their cots against the wall so that they looked like sofas. And they had their own living room at the end of the hall. But when they reconvened after a day with Lieutenant Mitchell and the other instructors, all they wanted to do was "hangar fly." Though they had plenty of coffee dates with officer friends, there was little time, or desire, to go to the base officers' club. They were in love with the Big Friend. The WASPs at Lockbourne now numbered fifteen. Two had left and were now stationed in Mississippi. By the end of December, two more would reach their limits in the B–17. "They didn't have Lieutenant Mitchell," Charlotte Mitchell would write home, lamenting their departure to her parents in Tulare, California. "Now we're the 'Lucky Thirteen.' " Grueling Fortress training was not for everyone. But for the "Lucky Thirteen" who remained, including all of Lieutenant Mitchell's students, the more they learned about the B–17, the more they believed they were flying the most wonderful airplane in the world.

BY MID-NOVEMBER, Lieutenant Mitchell's WASPs had over fifty hours in the Fortress. It was time for them to solo. As Major Fred Wilson, Lockbourne squadron commander, took them up one by one to check them out in the first pilot's seat, Lieutenant Mitchell spent all afternoon on the ground, pacing. Finally, Wilson walked into the ready room. He had just soloed Mary Parker, and he was beaming. Lieutenant Mitchell had done one hell of a job, he said. "I've rarely seen a student so proud of soloing that thing," Wilson added. "By the way, who's Charley Parker?"

"Her father, I think," Lieutenant Mitchell said. "Why?"

"We got out on the runway, were cleared for takeoff and just before she hit the throttle, she said, 'I'll show you, Charley Parker!' "

The two officers shook their heads. "These gals are something," Wilson said, and Lieutenant Mitchell had to agree.

OVER THE TWO YEARS he had been teaching B–17 students, Logue Mitchell had developed a style of instructing based on the theory that the more a student learned, the more his instructor should push him. Lieutenant Mitchell pushed hard. He had found that the more he expected and demanded, the more his students produced, as if they were trying to compete and keep up with him.

The WASPs were different. They were trying to keep up with some ideal they had to prove for themselves, and when he interfered, they let him know. A month after she soloed the B–17, Mary Parker was under the hood practicing navigation by radio beam. On the ground were radio transmitters assigned specific frequencies, corresponding with the points marked on air charts. To a pilot heading straight to or from a transmitter on a course of due north, east, south or west, the beacon broadcast a steady, humming signal into her earphones. Directly over the transmitter was a cone of silence. When the hum stopped, then resumed, the pilot knew exactly where he or she was on the map.

Lieutenant Mitchell, who had served from 1940 until Pearl Harbor as a staff pilot with the Canadian Royal Air Force and had flown hundreds of hours across uninhabited parts of Canada, issued adamant instructions to his WASP students on how to tune in a navigational radio. He insisted that, once on course, they continually turn the volume down as the open-range station was approached, to sharpen the pilots' ears. For a Fortress pilot flying over a station at 170 miles per hour, the cone of silence was just a second-long interruption in the signal, and easy to miss.

The steady signal was broadcast only within three degrees of 0, 90, 180 and 270 degrees. Keeping on course was called "bracketing the beam." Corrections to maintain the course, especially within ten miles of the station, had to be immediate and delicate, or the plane would drift into the transmission quadrant to the left or right of the hum. The steady hum would fade into Morse code beeps: an A, or "Da-dit" to the left; an N, or "Dit-da" to the right. The concentration necessary to keep the plane on course was worth the necessary effort. Drifting back and forth, as the hum broke into beeps and then recovered, only to disperse into other beeps, could drive a pilot wacky.

Mary had successfully bracketed the beam to two radio beacons. As she was the last to go under the hood, Lieutenant Mitchell instructed her to fly the plane home for lunch. She knew there was an open-range station eleven miles north of Lockbourne, and that if she headed for it, she could follow that compass heading and land right on the active runway. She got the hum into her earphones and followed it south. As she held the Fortress steadily on course, the hum got louder. She was so determined not to stray into the A or N quadrant that she gripped the wheel harder and forgot to turn down the volume on her radio. Just as she was almost over the station, however, she heard a confusion of Morse code and hum and soon realized that she had missed the cone of silence. Suddenly, Lieutenant Mitchell's head appeared next to hers under the hooded curtain. He was raging. "If you can't keep that volume down, you can't do a thing right!" he shouted. "You've got it up so damn loud you're through it before you know there's a change! It's gotta be soft, so your ear will be tuned to hear nothing!"

Greenie, standing in the top turret behind Mary, cringed. Then she heard a firm female voice resound from under the hood: "If you let me do that again, I'll show you I can do it!" Lieutenant Mitchell's head popped out from the curtain and he lurched back in his seat as if thrown. Greenie stared wide-eyed, fearing a violent reaction from the vehement young instructor. Instead, he was biting his lip to keep from laughing. Looking back at Greenie, his finger to his lips, he growled, "Time has run out, WASP Parker, I don't have time to fool around with you any more." Mustering his composure, he ripped the hood off the windscreen. Mary, face crunched with determination, flew them back to Lockbourne and made a perfect three-point landing. The next week she was given several chances to do it again, and she never missed a station.

After all of his yelling and griping, however, Lieutenant Mitchell could not figure out Charlotte Mitchell. He even began to lose sleep over her. The six-foot easygoing Californian puzzled him more than any student he had ever taught. For a week, she could do everything right. Then all of a sudden, she went into a fog, and for the rest of the morning had to ask, "What do I do now?" She would not know her number one engine from her number four.

For two weeks, he watched Charlotte like an OSS spy. He listened to her conversations with other WASPs and how quickly she answered questions. He watched the way her hips moved when she walked, and how she put on her jacket. One day, he was watching her dealing out playing cards to some other pilots in the lounge. One of the cards flipped out of her hand. In the fastest movement he had ever seen, Charlotte caught the card before it was halfway to the floor, stuck it back in the deck and resumed dealing. Well, he thought, chuckling in disbelief, it's not because she's uncoordinated.

Two days later, he sent Charlotte and the others out to do their routine walk-around and cockpit check. He watched out the window and waited for someone to warm up his copilot's seat for him. Suddenly he saw a flash of flame burst out of number two engine. Never before had anything gone wrong in any of the WASP planes. He dropped his coffee, raced to the Fortress and practically soared through the fuselage into the cockpit. "What happened?" he shouted. Charlotte was sitting in the pilot's seat. "Well, you told us if we have an engine fire we should feather the prop, so I did," she said, her blue eyes bright and earnest, fixed on his. He looked beyond her out to the wing. The fire was out. Lieutenant Mitchell patted her on the shoulder. "Beautiful work," he said. He would never worry about Charlotte again. Then he realized that he had rarely, if ever, complimented the tall affable young WASP, or any of his students. Charlotte's reaction to his insistent bullying, he thought to himself, was probably to shut him, the airplane and all that she knew, completely out of her mind. The young lieutenant realized he had to use other tactics as well, if he was to get the best out of his students.

Just before Christmas, Lieutenant Mitchell's wife, Lois, suggested he invite his WASP students to dinner. She had taken flying lessons before the war, and was curious about the women who were getting the chance to learn to fly the B–17.

As the WASPs entered the small, off-base apartment of their instructor, Mary Parker handed him a record album wrapped in tissue paper. Opening it, Lieutenant Mitchell found a recording of Strauss waltzes. He loved them, but he had no idea how they knew. The lanky young lieutenant looked down at the floor. For two years he had been flying 160 hours a month, seven days a week, turning out class after class of B–17 pilots. He had never before received a present. As he looked up at his WASP students, his blue eyes were shining. "This is the only time any of my students has ever given me anything," he said quietly.

For several moments, the WASPs shuffled their feet awkwardly in the middle of the living room. Then Mary blurted out, "I can understand why. They're so glad to get rid of you they'd never think of giving you a present!" Logue Mitchell's eyes rose to the ceiling. Communication was back to normal.

Through their stay at Lockbourne, the base medical staff watched the WASPs so closely that they even monitored their eating habits. One day in December, Lucille Friesen got orders to report to the infirmary. "You are not eating as well as you were," the flight surgeon said to the five-foot-ten-inch WASP from Kansas. "Is everything all right?" Lucille assured him she was feeling fine. The dreamy-eyed brunette was in love.

The few occasions they had the time and the energy to go to the officers' club, the WASPs were always a hit, in spite of their rumpled slacks. Official WASP uniforms would not be issued until spring, but none of the officers seemed to mind what they wore. One evening, Lucille was chatting with Lockbourne's Director of Flying, Captain John McVey, and he suggested they have a cocktail. He offered her a French 75, a popular drink around the club. A highball mixture of gin and champagne over ice, it was aptly named after a French field artillery gun. WASP Friesen and Captain McVey, soon enraptured, talked about flying all evening.

On night cross-country rides, Captain McVey would designate Lucille's as a check ride so that he could go along. She sat in the pilot's seat, another WASP in the copilot's seat, and Captain McVey stood in between, watching the gauges. The calmest of pilots, Lucille did beautifully, even though every once in a while, Captain McVey leaned over and gave her a kiss on the cheek. The flight engineer sitting behind them grumbled, "Gee, I wish I was a Director of Flying. . . ."

One evening around Christmas, Lucille came back late to the nurses' quarters after a late date with Captain McVey. Mary Parker was still awake and reading in the lounge. "Now, Lu," she said, "you've been seeing this guy day and night, why don't you marry him?" "I'm going to!" answered Lucille. The Director of Flying had just proposed.

Outfitting a bride was not easy in the winter of 1943. Mary spent several afternoons combing Columbus for a dress suitable for her friend to be married in. Three weeks later, she had miraculously found not only a dress, but a hat and a pair of silk stockings. She was yet to top off her trousseau, however. One evening, Lucille arrived home to find, lying on her bed, a satin nightgown. She rushed to Mary's room. "You shouldn't have!" she said breathlessly. "For heaven's sake," retorted the sardonic upstate New Yorker, "you've got to look pretty on your wedding night." Lucille blushed.

After three weeks of flying near Lockbourne, Lieutenant Mitchell began to take his flight farther afield, to teach them cross-country navigation. Now they were flying mornings and evenings. He would send four of his students into the plexiglass nose and gun turrets to watch for check points and beacons, put one in the radio compartment, while the sixth flew as pilot.

Now that the WASPs were coming into contact with the world outside of Lockbourne where they were a familiar sight, the women B–17 pilots were reminded what an oddity they were.

On one cross-country flight at night, at 8,000 feet above Virginia, Lieutenant Mitchell asked Mary Parker, stationed in the radio compartment, to make a position report. She dialed in the Roanoke frequency and called the tower. "This is B–17 number 137, do you read?" There was no answer. It was ten o'clock and she figured the radio operator was dozing. She called again. Suddenly there was static and a voice shouted, "B–17 137, what are you doing up there?" Mary said, "I'm flying a B–17 and I would like to make a position report, do you read?" "Roger, I can read," said the operator. "Give me your position report." The news of a woman's voice coming from a Flying Fortress woke up the entire airport.

On another cross-country, when they emerged from the B–17, they grabbed Lieutenant Mitchell by the arms and made him walk into the operations building with all six of them. They were not sure that they would be welcomed as legitimate pilots. They also enjoyed flustering the usually stern young lieutenant, who, accompanied by his sensational bevy, was met with open-mouthed stares by operations officers.

AFTER SOLOING, the WASPs were now entrusted with their own Flying Fortresses, so that they could practice flying cross-country as operational pilots. Since their arrival on October 15, they had flown the Fortress for seventy-five hours, but they still had fifty-five more to go, and most of them would be on trips.

One afternoon in early December, Mary Parker and Helen Dettweiler were given orders to fly a Fortress to Detroit to take an officer to see a sick aunt. The only male crew would be a navigator. Helen and Mary stood in the operations office laughing as several potential candidates flipped a coin to win a flight with the WASPs. The once reluctant crews now viewed a flight with the celebrated women pilots as a prized event.

On one day-long cross-country over the midatlantic states, four WASPs and four men B–17 students learned via radio reports that their home base was socked in by a storm, and would be for at least twelve to fourteen hours. There was no choice but to land at the nearest open field, Pittsburgh.

From the airport, Helen Dettweiler called the manager of the Pittsburgher Hotel, whom she knew from various golf tours before the war. "We're stranded because of the weather," she explained. "Do you have any rooms? We don't have any clothes with us but our flight suits." He said he would make room.

The first night, they played cards and had food sent up by room service because they did not want to go into the lobby in their flying suits. But the weather settled in for one, then two, and finally five full days. On the third night, they got tired of playing cards and looking at each other. By asking the hotel help, they learned that there was a nightclub a block away that had a floor show and where any dress was appropriate.

In minutes the eight pilots were walking down the street in their zip-up flying suits. People stared at the bizarre group of men and women in baggy overalls. When they reached the address, they went inside; soon the lights dimmed and the show began. The WASPs stared. Onto the stage came several women dancers, about their age, with nothing on but G-strings and tiny propellers fastened to their nipples. As the dancers began to gyrate, the propellers twirled in opposite directions. Thematically the two groups of women had something in common, but not much else.

FINAL FLIGHT CHECKS were scheduled for mid-January. For several weeks, however, the weather over Lockbourne had been undependable. Either snow was falling or smoke and haze from the coal-burning furnaces of Columbus sailed over the field.

One of the days they were grounded, they reported to the Quartermaster's to be measured for the official WASP uniforms. Unlike other fliers in the Army Air Forces who wore khaki, the WASPs would wear uniforms of a lively Santiago blue: belted "Eisenhower" jackets, shirts and slacks. The WASP topcoats, they were glad to hear, would be issued soon. They left the Quartermaster's in a blizzard.

Several of the pilots could not get away for Christmas. They stayed in the nurses' quarters, listened to Bing Crosby sing "White Christmas" on the radio and exchanged presents. The WASPs gave one another perfume, hosiery, and subscriptions to *Reader's Digest*. Charlotte Mitchell got a wool sweater with P–38s and little parachutes knit all over the front. Mary Parker gave her a silver crash bracelet with her name and serial number engraved on it. But the high point of the Christmas season was when Squadron Commander of the 1174th, Major Fred Wilson, gave all the WASPs gifts. On each was a tag attached with a silver star, which read "Happy landings for ________ From Fred." Surprised, they opened the presents carefully. Their dashing squadron commander had given them all silver dollars with miniature AAF wings soldered to them. The WASPs were speechless. Pilots always had a superstitious streak, and that

Major Wilson had given them such beautiful lucky pieces moved them deeply.*

By New Year's Day, however, the WASPs had thirty hours of cross-country flying still to go. As day after day of low ceilings kept them on the ground, Lieutenant Mitchell decided to take his WASPs on a long cross-country trip, to Houston, Texas, 1,200 miles to the south, a point not plagued by storms. They simply had to get sufficient air time before they left Lockbourne. To make sure they would have calmer skies, they decided to fly at night.

On the evening of takeoff, Greenie put on her long underwear, zipped up her leather fleece-lined pants and hooked her suspenders, put on her fleece-lined boots, and finally zipped up the bulky leather jacket. She joined Mary, Charlotte and the others. They all walked across the sleeping base to meet Lieutenant Mitchell and their crew chief by their olive-green Fortress. Light snow was falling. "The first hour might be totally by instruments," Lieutenant Mitchell said. "You gals are rated for it, and we have to get above the storm." He seemed more grim than usual and was making an effort to be lighthearted.

He asked Mary to take off. "It's just like a hooded takeoff," she said.

"That was the whole point of doing them," Lieutenant Mitchell responded. "This one's for real."

They taxied out, and when they were exactly on the runway heading, he locked the tail wheel. Slowly Mary rocked the throttles forward, built up speed and finally the Fortress broke ground. One hundred feet up, they could no longer make out any of Lockbourne's buildings. Mary sat riveted to the gyro compass and airspeed indicator, as she had been trained, and at pattern altitude made a precise climbing turn in the direction of Houston. They had a good ten hours of air time ahead.

As they climbed through the muck, Mary did not dare look out the windscreen into the gray mass of air or she would get completely disoriented. The sky looked like the murkey depths of the ocean. She almost expected to see fish. As they climbed, the temperature dropped. At 10,000 feet they all put on oxygen masks. As they climbed through 24,000 feet, the thermometer read twenty degrees below zero.

* Twenty-four years later, Virginia Acher Williams would give her silver dollar to her son Richard, a helicopter pilot, before he left for Vietnam in 1967. He returned home safely to Billerica, Massachusetts, but he had lost Major Fred Wilson's lucky piece.

"The glycol heating's rationed in this plane," joked Lieutenant Mitchell, but the ice crystals on the edges of his breath, like those forming on the windshield, belied the humor in his remark.

The WASPs standing behind Mary and Logue Mitchell began dancing like cartoon sausages, from one foot to the other. Suddenly there was a deafening crash. Mary's hands jumped off the control wheel. The noise sounded as if they had just been hit by a .90-caliber shell. "What was that?" shouted the WASPs in unison. Lieutenant Mitchell attempted a nonchalant shrug. "Just ice forming on the props," he said. "When it breaks off it hits the side of the plane." Just at that moment several more sharp crashes sounded throughout the plane, and their hearts quickened and pounded.

At 35,000 feet, Mary let out a whoop, as they at last broke through the clouds. Suddenly they were in a black sky and gazing ahead of the Fortress over endless cloud hills glowing eerily in the moonlight. Forgetting their cold, everyone looked in awe as Mary leveled off. None of them had ever seen the world from such a height. "Look at our airplane!" Greenie shouted. They looked out at the wings and what had been a drab olive Army green was now gleaming aluminum, stripped of paint by the ice banging against the metal.

To pass the time and keep up some movement in the glacial cockpit, Lieutenant Mitchell had everyone rotate positions as pilot, copilot and navigator. He asked constantly for position checks.

Halfway across Tennessee, Greenie appeared nervously behind Lieutenant Mitchell. She had to urinate. In the Flying Fortress, facilities were not deluxe. Through the radio area, past the bomb bays and back into the tail gunner's station was a bottle and a funnel. Under 10,000 feet, when supplemental oxygen was not needed, getting back and forth to the funnel was no problem. Now at 40,000 feet, Greenie would have to stray far from the central cockpit oxygen system into which their masks were plugged. Lieutenant Mitchell fitted Greenie with a mask attached to a walk-around bottle containing enough oxygen for three to five minutes of normal breathing. She clamped the oxygen mask around her head, smiled uncertainly and left the cockpit. As she closed the cabin door, Lieutenant Mitchell clicked his stopwatch.

By the time she had clamored back through the fuselage, unharnessed her parachute, unzipped and taken off her jacket, unzipped her fleece-lined pants, pulled the suspenders off her shoulders and finally got down to her long underwear, she was panting with exertion and shivering almost uncontrollably. She managed to squat over the funnel. Then she pulled up her long underwear, her pants, and slipped the suspenders back over her shoulders. But as she tried to zip her leather pants up the leg, the zipper got caught in her long underwear. For several seconds, her teeth chattering and her hands numb, she tugged at the zipper.

The next thing she knew, the crew chief was holding an oxygen mask to her face, and she was slumped on the floor. "Hurry up, let's get back up there," he said brusquely, trying to hide his embarrassment. "You were almost a goner." He gave her a new oxygen bottle and disappeared while she finished zipping herself and pulling on her jacket and parachute harness. When she finally opened the cockpit door, everyone was so relieved that she decided not to make a joke—like the one about the long lines in the ladies' room at this joint—and just grinned sheepishly.

Over Shreveport, Louisiana, the sky broke under them and they began their descent toward Houston. As they went down, the cockpit became warmer. When it got above freezing, they began to unzip their sweaty flight suits. By the time they landed in Houston, the sky was clear and the temperature sixty-five degrees. One by one they climbed down from their shiny Flying Fortress. Greenie dropped to her knees and kissed the ground. "Texas!" she shouted. "What did I tell ya? It's the greatest state in the Union." For once, they all had to agree. Even Lieutenant Mitchell leaned down and patted the warm earth gratefully.

Over the next three days, from sunup to sundown, the WASPs mounted up almost half of the remaining time they needed to graduate. They also ate what seemed to be half the steaks in the city of Houston.

When they flew back to Lockbourne, the storm had spent itself and the return trip was breathtakingly lovely. They passed over almost 1,000 miles of gleaming, silent, snowbound earth between Oklahoma and Ohio. With no aerial ice to battle, Lieutenant Mitchell and his students flew relaxed. Along with the first pilot's seat, the prize position was navigator in the nose with nothing but a clear barrier of plexiglass between her and the world spread out before her.

Back at Lockbourne, they were met by Lucille Friesen. She had been flying continuously since the storm had passed to get in her airtime. But Lucille had become worried about the WASPs' return. It was now Thursday, and on Sunday, Greenie was to be her maid of honor.

The WASP wing of Center Hall was suddenly festive. The wedding was turning into the biggest social event that Lockbourne Army

Air Base had ever seen. Invitations had gone out by word of mouth around the base. Lucille had no idea how many would come.

At noon on Sunday, Lucille and her husband-to-be stood at the back of the base chapel facing a long red carpet to the altar. Almost a hundred officers, her fellow B–17 students and other air base personnel, packed the pews. Interspersed was the tiny squadron of WASPs. Lucille spotted Mary Parker in the middle of the room. As she waved, Lucille noticed Mary's eyes were already misty. "I'd rather be flying in a hurricane than this!" Lucille whispered to her future husband. Captain McVey patted her hand reassuringly.

Suddenly the organ struck up the wedding march. John had gone to the front of the church, and Major Fred Wilson appeared with a dazzling smile and offered his arm to Lucille. The handsome Squadron Commander of the 1174th training squadron would give the bride away to his Director of Flying. Suddenly WASP Lucille Friesen found herself walking down the aisle.

After the ceremony, Lucille and John McVey posed for pictures outside the base chapel. Photographers from the Chicago *Tribune*, the New York *Times* and the Associated Press had flown into Lockbourne Army Air Base to capture the first wedding of two B–17 Flying Fortress pilots in American history.

For the next two weeks, the WASPs flew all over Ohio and at last completed their 130 hours of flight training in the B–17 Flying Fortress. By now the lumbering bomber was indeed a friend. But between studenthood and graduation was one final check ride, the most grueling of tests of their flying ability in the four-engine Flying Fortress.

The morning of January 11, Charlotte Mitchell was pulling on her flying jacket in the ready room when Major Fred Wilson moseyed over to Lieutenant Mitchell and his group of WASP students. "I think I'll go for a ride with you this morning," he said casually, his handsome face breaking into a smile. "I haven't ridden with you yet, have I?" Charlotte immediately stiffened. The mere presence of the highly attractive young bachelor, who throughout their training had remained aloof, was heart-fluttering in itself. But now he was going to judge her on everything she knew about the airplane she had lived in since October 15, for almost three months. Charlotte did not notice, but Lieutenant Mitchell became just as nervous as she.

Charlotte, Lieutenant Mitchell and Major Wilson boarded a B–17 awaiting them out on the ramp. Charlotte sat in the left seat, Lieutenant Mitchell in the copilot's seat, and the squadron C.O. in a jump seat between them. Charlotte fingered the silver dollar in the pocket of her flight jacket, then she gave herself over to the check ride. "Let's do a couple of hooded takeoffs," suggested Major Wilson cheerily. As Lieutenant Mitchell began to sweat, Major Wilson pulled the curtain over Charlotte's face. Charlotte caught a whiff of the soap he shaved with.

After all of Lieutenant Mitchell's patient repetition, her hooded takeoffs were perfect. Once airborne, the curtain was pulled back and she was made to fly with Major Wilson's hand blocking instrument after instrument. Then the major told her to bracket the beam toward a radio beacon, and then take coordinates on several others to pinpoint her exact location. Lieutenant Mitchell gave her frantic hand signals each time Major Wilson coughed or sneezed. Her instructor's furrowed brow and wild blue eyes made Charlotte want to laugh. She realized that Lieutenant Mitchell wanted her to do well as much as she wanted to herself. But just as she was gaining some perspective on the flight check, Major Wilson put his chin on her shoulder.

After about an hour of various tests, Major Wilson ordered her to climb to 25,000 feet for some high-altitude maneuvers. They put on their oxygen masks and relaxed for twenty minutes as the Fortress plowed up through the sky. When they reached 25,000, Charlotte executed some stalls and turns in the low-pressure atmosphere. All was going well.

Finally, they descended toward Lockbourne. After two touch-and-go landings, two hours had gone by. She taxied back to the hangar and as they climbed down out of the Fortress, Lieutenant Mitchell and Major Wilson were slapping her on the back and congratulating her with ebullience. Charlotte could only manage a weak smile. She felt like a wrung-out rag.

ON JANUARY 15, the officers' club at the Fort Hayes Hotel in Columbus, Ohio, was ablaze with lights. A twelve-piece orchestra was playing swing tunes and a crowd of dancers bobbed around the floor. Major Fred Wilson, Lieutenant Mitchell and the other flight instructors had organized a gala dinner dance to celebrate the graduation of thirteen WASPs from B-17 transition school. Major Wilson contributed bottles of champagne for every table.

Graduation from B-17 school usually consisted of a check ride and the signing of a certificate. But Lockbourne could not let their wonderful ladies of the Fortress go without showing them how proud they were of them. And the WASPs were getting a measure of just how much stamina they had built up over the rigorous three months at Lockbourne. Their dance cards were filled with names of officers, some of them their fellow students, who were also through with their training but had never expected to celebrate in such elegant style.

The triumphant women pilots were giddy with congratulations from their dance partners. To their delight, even Major Wilson danced with every one of his successful check rides. But for six WASPs, their favorite partner, the one whose words meant the most, was the tall gangly lieutenant named Logue Mitchell. He had promised to get everyone of them through and all six of his students had passed with flying colors. It felt strange to be holding their flight instructor, who had barked, taunted and pushed them into sweats of exhaustion, in their arms, but dancing finally allowed them to give the young lieutenant a warm hug of gratitude.

The orchestra played until the early morning, but when the strains of "Good Night, Ladies" finally filled the room, the victorious ladies of Lockbourne felt that they could dance on for many days to come.

XII

Women Under Fire

While Kay Menges was having her new parachute fitted in the flight operations office of New Castle Army Air Base, a classmate from 43–3 rushed in shouting, "Don't do it! We're being reassigned!" In her hand were Kay's orders, dated July 14, 1943. Though they had been at New Castle only for a couple of days, they and twenty-three others who had graduated on July 3 from Avenger Field were to report in five days to Jacqueline Cochran at the Mayflower Hotel, Washington, D.C.

On Monday, July 19, Kay found herself in the swank lobby of the Mayflower on Connecticut Avenue in the nation's capital. Neither she nor any of her classmates knew why they were there, and as Jacqueline Cochran gathered them up for an elegant dinner at the hotel, the aura of mystery grew. They had been selected for a top-secret experiment, Cochran told them. General H. H. "Hap" Arnold himself would tell them what it was.

The 43–3 graduates were taken in buses to Bolling Field in nearby Virginia to be tested in a high-altitude chamber and certified to fly as high as 38,000 feet. They did sightseeing in Washington and were wined and dined like visiting dignitaries.

Finally, buses drove them from the Mayflower to the newly completed Pentagon to meet with the Commanding General of the Army Air Forces. After being given clearance badges, they were led through long Pentagon corridors, lined with portraits of generals and paintings of airplanes, and into a large conference room. The entire room seemed to be padded with red leather. Cochran seated the women pilots around a huge wooden conference table. After waiting for several minutes in respectful silence, they suddenly rose from their chairs when a tall, solidly built man with an irresistible smile strode into the room—General "Hap" Arnold. He greeted them all, congratulated them for being selected for this very special program in which they were about to participate, and then introduced Jacqueline Cochran, Director of Women Pilots. Impressed by the formality with which they were being treated, the graduates of 43–3 thought the program must be very special indeed.

Cochran got up from her chair, put two palms flat on the table and leaned forward, her large brown eyes surveying the faces of her graduates. "You are going to be assigned to a top-secret mission," she said. "You are going to have a chance to fly bigger and better airplanes than women have ever flown before. Is there anyone who doesn't want to be assigned to this mission?" Though no one knew what the mission was, there were no objections.

Then Cochran explained that for the first time women pilots would be assigned to flying duties other than ferrying airplanes. They were about to be given the opportunity to serve in one of the most crucial and central domestic functions performed by pilots of the AAF. Their performance on this mission would determine the future of women pilots in many other capacities. The conference went on for fifteen more minutes and though the excitement mounted around the room, no specifics emerged about their mission. Then they were dismissed.

The following day, after a final elaborate dinner, the graduates were packed aboard a bus to an air force base outside of Washington where they boarded a beautiful DC–3 military airliner. Some believed it was General Arnold's personal airplane. Two male pilots, high-ranking officers, climbed into the cockpit, and the plane took off toward the south. As they flew over the rolling green Piedmont plains of Virginia, the sun set. Shortly after dusk, they landed; the two officers left the plane and went into the airport operations office. Judging by their concerned looks, the women got the distinct impression that their male counterparts were lost. Fresh from advanced cross-country training to prepare them for ferrying duty, the 43–3 graduates suppressed smiles as their DC–3 pilots climbed back through the cabin. Soon the airplane was heading down the North Carolina coast, over the Outer Banks where Wilber and Orville Wright had flown their *Flyer* forty years before, and then descended to a few hundred feet over the darkened beaches. Nestled in the swamps just inland from the town of Wrightsville Beach, they spotted the barest outlines of an airfield. "That's it," one of the pilots shouted back to the cabin, "Camp Davis."

WHEN Jacqueline Cochran was named Director of Women Pilots in the AAF, and assistant to the Chief of Air Staff for Commitments and Recruitment on June 28, 1943, she knew immediately what she wanted to do. Now that her training school graduates had been successfully flying for the Air Transport Command since mid-May, her first priority was to expand the number and types of assignments available to women pilots. One of the largest domestic flying operations was to assist in the training of thousands of AAF air-to-air and ground-to-air gunners. The squadrons towing targets at gunnery training bases were so badly in need of pilots that the Third Air Force had recently transferred an entire graduating class of combat-bound cadets to perform tow-target missions at Camp Davis, North Carolina. Men pilots agitating to go overseas to fly against the enemy and yearning for the glories of acehood considered such domestic assignments as a crushing disappointment, if not a humiliation. If women could take over the towing of targets, then able-bodied, flight-trained young men could be released for combat duty. Cochran proposed that some of her training school graduates be assigned as an experimental group to the tow-target squadron at Camp Davis. General Arnold approved the proposal.

The second week on her new job, Cochran went into action. First of all, she requested of the Air Transport Command that twenty-five of her recent graduates be ordered for temporary duty to Washington. Then she wrote a confidential letter to Major Robert K. Urban, commanding officer of Avenger Field in Sweetwater. Since tow-target duty would entail high-altitude flying in such airplanes as the B–26 and B–24, she requested that Major Urban recommend a list of twenty-five of the best pilots of the current class of 43–4, who would be graduating on August 7. These pilots had to be tops, she wrote Urban, "as they will be setting the pace for many hundreds more to be used in this capacity if it works out." On August 6, twenty-five 43–4 graduates would receive their postgraduation orders —to report on August 20 not to a ferry base, as they expected, but to Jacqueline Cochran, Room 4D957, The Pentagon. They would then join the first group of women to make up a special squadron of fifty WASP pilots.

* * *

Camp Davis, one of the AAF's largest and oldest antiaircraft artillery training bases, was situated on the edge of North Carolina's Dismal Swamp. Stationed there were 50,000 officers and enlisted men, whose purpose was teaching and learning to operate and maintain all types of antiaircraft guns, from rifles to towering .50- and .90-caliber artillery.

As support, 600 pilots were also stationed at Camp Davis with a sole mission—to act in the air as much like the enemy as possible. The gunnery trainees attending the Anti-Aircraft Command school at Camp Davis were there to learn how to hit their targets—which upon graduation, would be Axis aircraft in Africa and the Pacific. The pilots flew prescribed patterns above gun emplacements at several sites around the huge tract of Camp Davis. Missions were also flown at night, when airplanes would be tracked by searchlights. The largest gunnery range was along Wrightsville Beach. Pilots flew, dragging twenty-foot muslin sleeve targets behind them, from one end of the beach to the other, as one by one each gunner took aim and fired, first tracer bullets and then live ammunition, at the sleeve targets.

When the news began circulating around Camp Davis that two dozen young women pilots were about to arrive, a group of enlisted men went to the tow-target squadron commanding officer, Major L. L. Stevenson, and requested transfers. "I'm not going to serve any powder puff pilots," grumbled one mechanic. Major Stevenson asked them to stay until he could see if the women worked out. Then if the men were still disgruntled, he would arrange for their transfers. Concurrently, subdued mutterings among the male pilots hinted at a strike if women were allowed to do the same work they did, further degrading their noncombat role.

Their first morning at Camp Davis, the women pilots awoke with eager anticipation about their top-secret assignment. The roar of airplanes met their ears. As they looked out the windows of the nurses' barracks where they had been billetted the night before, they saw that their new quarters were situated directly on the edge of one of the runways.

After breakfast in the officers' mess, they were met at the flightline by Major Stevenson, a short, stocky, balding man around fifty years old, who spoke in a thick Southern accent. He led the group of women, still aglow from the regal treatment and dramatic build-up given them in Washington, past the A–24 and A–25 dive bombers, the twin-engine B–34 bombers and the C–45 transports. Finally, they reached a line of L4 and L5 Cubs. "You will check out on these today and then maybe fly them on administrative missions," he said. The women pilots stared. Some looked confused. Kay Menges had been in a Piper Cub only a couple of times in her life, for her first flying lessons at Teterboro Airport, New Jersey; on both occasions, her flight instructor had become airsick. Others in the group, who felt they had left Cubs far behind, tried not to appear indignant.

As they took off in the little airplanes, they began to feel better. They would show the skeptical commanding officer that at least they could fly. As the first few WASPs came in for their landings, however, their attitude turned to consternation. Some of their airplanes had heel brakes.

Kay Menges and the other women pilots had just completed 200 hours in Army trainers in which the brakes, attached atop the rudder pedals, were pressed with their toes to slow down the airplane after landing, and to taxi. They had operated the rudder pedals with their heels, to steer the plane just before and after landing. In the Cubs, their instincts were all wrong.

As Kay prepared to touch down, she automatically settled her feet flat on the rudder pedals to get ready to hold the plane straight as she sped down the runway. As soon as she hit the runway, however, her feet on the pedals not only moved the rudder but her heels activated the brakes. The plane jolted forward and almost nosed over, until she realized her mistake.

All afternoon, the newly arrived women pilots landed in the ignominious little Cubs and screeched and jerked down the runway and over the taxiways. The men pilots on the flightline, their expectations met, laughed uproariously. Jacqueline Cochran's finest, so eager to be a successful experiment, were totally exasperated.

The Director of Women Pilots soon learned that her experiment, once out of the planning stages at the Pentagon, was not working very well. Three days later, Cochran flew to Camp Davis to acquaint herself with the tow-target operation for which she had volunteered her women, and to have a conference with their new commanding officer, Major Stevenson. When she saw what her prize graduates were doing, she sensed trouble. "I didn't train these girls for six months as multi-engine-rated pilots just to be flying around in Cubs," she protested. Finally Major Stevenson relented. Against his better judgment, he promised to check them out in the A–24 Douglas Dauntless, a tow-target ship.

Being a mere twenty-five women on a base of tens of thousands had its hazards. Dates were available any time, but in their first few weeks, the WASPs were much too concerned holding their own as professionals in the tough-minded squadron of pilots to carouse with AAF officers. In fact, the amount of drinking among their pilot colleagues was shocking to some of the WASPs, who needed every ounce of energy and concentration not to fail in any aspect of their special assignment. Every once in a while, however, a WASP or two would acquiesce to a poignant plea at their barracks door and nurse a drink for a while, laugh at the placemats at the officers' club, which were covered with cartoons of all the Axis dictators, and have a dance or two. But soon they would find themselves so droopy-eyed with the tension and newness of their flying duties that they would admit they were bushed and go home to their cots for a deep and sweaty midsummer sleep. Any worries Jacqueline Cochran or their parents might have had about a small select crew of women on an air base with a hundred-to-one ratio would have been defrayed after one look at the exhausted young women pilots who, if a man gave her more trouble than sympathy, would just as soon be in bed—and alone—than out galavanting. Nevertheless, the WASPs were so beleaguered that Camp Davis quickly became known around the women pilots' barracks as "Wolf Swamp." This epithet soon extended to official air base personnel.

Two weeks after her first visit to Camp Davis, Jacqueline Cochran was back, this time to speak with the base medical officer. He had announced that every month complete physical examinations would be required of all women pilots. An apprehensive 43–3 graduate had telephoned Cochran in Washington. "You don't examine men pilots once a month," Cochran exclaimed to the base medical staff. "No one is going to examine my girls unless I authorize him to." The routine physical examinations were canceled.

Shortly after Cochran's protest to Major Stevenson, he allowed his new WASPs to check out in the 1,000-horsepower Douglas Dauntless dive bomber. His eagerness for this enterprise had not grown any keener. "These planes are dispensable," he told the women pilots as if he were sending them off on a bombing raid over Berlin, "and *you're* dispensable." The women pilots did not know whether or not to laugh at his joke.

The transition from a 650-horsepower AT-6 to the Douglas Dauntless was not a major step up the aerodynamic chart, but after such reluctance on the part of high command, the women were thrown into the cockpit alone at the controls of the Dauntless with surprising nonchalance. Their check-out consisted of one ride in the back, or gunner's, seat of the tandem cockpit of the plane. The ride was not particularly instructive. The rear cockpit did not have working gauges, so they had to guess what the check pilot was doing by sound.

The Dauntless, primarily used by the Navy, had hydraulically operated perforated dive flaps which slipped out beyond the trailing edge of the wings and acted as dive brakes for quick landings on aircraft carriers. The plane was flown, nose down, almost into the runway; then the pilot pulled up the nose suddenly, just before the wheels hit the ground. Trained as Army ferry pilots, the women had never flown planes with dive flaps, nor ridden as passengers in them. Having no warning of this jolt, they hit their heads on the metal gunsights. After their training flights, they all gathered on the flightline and looked at one another a bit dazed. Then they were handed tech orders describing the operation of the airplane and told to take an A–24 themselves.

Twenty-three-year-old Marion Hanrahan, who had been flying for nine years at Bendix Field, New Jersey, skimmed the tech orders and surveyed the cockpit of the Dauntless. As there seemed to be no instructor pilots around to answer questions, she grabbed a passing mechanic. Startled at being consulted, he explained to her the instruments, flight characteristics, and some of the idiosyncrasies of the plane. Marion held up a plastic packet from a side pocket in the cockpit. "What is this?" she asked. The mechanic, suddenly flustered, muttered, "You don't need that, put it back." "But what is it?" Marion asked again. The mechanic thought a moment." It's a pressure release valve," he answered finally. Several days later, Marion learned that AAF male pilots' planes were outfitted with relief-tube urinals.

Faced with the indifference of the pilot squadron instructors, as evidenced by the perfunctory check-out they had received in the Dauntless, the women quickly realized that the Camp Davis mechanics were going to be indispensable allies. At Sweetwater, they had been taught always to check Form One in the cockpit to verify each plane's airworthiness or state of repair. After a flight, a pilot wrote down on Form One anything he or she thought was amiss. If mechanics fixed what was wrong, the repairs were indicated on Form One. If mechanics determined at any point that the aircraft was in such need of work that it was unfit to fly, they marked a red "X" on Form One. But there was also a marking which was half an X, or a diagonal red line, which meant that the plane might have any number of things wrong with it but, technically, it could be flown in an emergency. Sometimes the airplane was barely flyable—the wings and tail were still attached, but some of the instruments were nonfunctioning and the radio did not work properly. In pilots' jargon, such a plane was said to be "red-lined."

The women pilots were instructed to fly the A–24s solo for a few hours before they would be taught how to tow "sleeves," or targets, for gunnery training missions. They soon learned that many of the A–24s at Camp Davis were red-lined. One by one, they began to experience many hazards which were expected by World War I Aces, but with aerodynamic advances of the twenty intervening years, were infrequent and dreaded by most pilots in 1943.

After one engine failure, from which she could barely glide safely back to the runway, and two blowouts, Marion Hanrahan began to observe the mechanics at Camp Davis closely. Marion had been maintaining airplanes ever since she was fourteen, when she first paid for flying lessons by patching wing fabric with sheets from her mother's linen closet. Some of the men, whom she determined must have been in training, knew less about engines than she did. Others seemed to have grown indifferent in the face of persistent shortages of parts. Two particularly spirited young mechanics, however, Marion decided to befriend. As they got to know her, they began to reveal the hardships under which they were working, as well as the mishaps she could expect while flying at Camp Davis.

Blowouts were almost inevitable in the old, worn tires on the A–24 Dauntlesses and A–25s, the faster, heavier Curtiss Helldivers. Every possible rubber airplane tire was being shipped overseas to combat zones. The instruments could not be trusted either, as overworked mechanics chose to spend most of their time maintaining engines so tow-target missions at least could be flown. Such incidentals as flap handles that would not lock into position they considered not even worthy of attention.

As Camp Davis was primarily an antiaircraft artillery training base, and not for training pilots, flying operations were short-shrifted all down the line. Replacement parts, such as spark plugs, ran out constantly. But proper engine maintenance was ultimately hampered by Camp Davis' fuel allocation, which was so small compared to the prodigious number of hours flown that the fuel tanks of A–24s were often filled with a lower octane than their Wright 1820s required. Marion was beginning to understand why the C.O. called the planes dispensable.

One day, she started the engine of an A–24 and it ran so rough she taxied it back to the hangar to consult the mechanics she had befriended. Together they drained a cupful of water out of the carburetor. "This happens all the time," her friends told her; in fact, so often, that they sometimes suspected sabotage. Water in the gas could have been caused by not filling the tanks at the end of the day so that water vapor condensed and collected overnight, or by filling tanks from a nearly empty drum which was half condensed water, half gasoline. But the mechanics doubted that any flightline ground crew could be so consistently careless.

Sabotage or not, Marion did fear that the lackadaisical attitude she had seen all over Camp Davis had infected the ground crew as well. When she and the other WASPs began to fly tow-target missions, she arranged with her mechanics to go over every airplane of which she took the controls.

Back at the WASP barracks, conversation, which had been subdued as the WASPs felt their way through the first days of their special assignment, began to evidence increasing alarm at the missions and airplanes they were flying. After the eagerness of their first day, morale had steadily dropped. Marion and several others exchanged the fear that they were losing their nerve. They realized that their view of themselves on graduation day at Sweetwater, as Army pilots ferrying bombers across America for the war effort, had perhaps been idealistic. But this special assignment was no fun at all. "We're Jacqueline Cochran's guinea pigs," exclaimed one WASP who had barely made it back to base when her engine quit in midair halfway home from flying over Wrightsville Beach gunnery emplacements.

Their confidence was further eroded when they started to be ordered to fly with search-and-rescue missions when pilots from Camp Davis had forced landings in the swamps surrounding the base. Proper emergency procedures for ditching in the swamps, they were told, were to pancake the plane in for landing so that it cut a broad swath, which would be visible from the air, through the trees and swampgrass, and then spread out their parachute. Then the pilot was to stay with the plane. Somehow the pilots were always found, but the collection of insect bites and spooky tales that remained afterward gave some of the WASPs nightmares.

In spite of the condition of aircraft and flightline maintenance, mishaps were not greeted with any sympathy by base command. One WASP was taxiing her A–24 to the hangar when the massive wing of a B–34 bomber suddenly loomed in front of her eyes. She jammed on the brakes and instead of skidding to a halt, the plane froze on the spot and nosed over. The ground crew came rushing over and pulled her, unhurt but flustered with embarrassment, from the cockpit. She then saw that the B–34 had blown a tire and been left, its long wing jutting halfway across the taxiway. The brakes on her A–24 had just been adjusted, but the repair had not been marked on its Form One.

Nevertheless, she was called before the squadron commanding officer. "What's the matter with you, you flying for the Nazis?" the commanding officer roared. "That airplane cost me $60,000! I wouldn't have *my* daughter flying airplanes." The WASP felt mortified. She had only damaged the propeller, and it had been her first accident in over 400 hours of flying experience. She was given a chance to explain the circumstances of the mishap, but when he dismissed her, she expected from the vehemence of his reaction to be court-martialed. When several days later no action had been taken against her, she decided that he had merely been venting his displeasure at having the women pilots on his base, and stopped worrying that she had ruined her future with the AAF.

At last they were flying the missions Jacqueline Cochran had wanted them to, towing targets for antiaircraft artillery practice. Though when they had flown for the Civil Air Patrol before the WASPs, they had not been allowed to fly coast patrol because it was too dangerous, now they would fly missions up to fifty miles out over the Atlantic. They now faced the realities of training for a war.

After a briefing at flight operations, Marion Hanrahan and an enlisted man, who would ride in the back seat and operate the target sleeve and cable, climbed into a Douglas Dauntless. They took off and once Marion was over the beach, she radioed the artillery officer in charge on the ground, as the cable operator began to turn the winch handle and let out the cable to which the muslin sleeve target was attached. The sleeve could be unhooked by a lever near the gunner's seat. When they had flown a couple of gunnery patterns down the beach, Marion was to swoop down over the sand and the cable operator would release the sleeve from the cable so that the gunners could see evidence of their accuracy.

Marion looked down the beachhead. Extending for about a mile were positions for batteries of antiaircraft guns. At one end of the beach were the three-inch guns, then .40- and .35-millimeter guns, and finally, down at the end, small-arms trainees with rifles. She would start at one end and fly at the altitude given her by the gunnery officer, then as she got to the small-arms range, she would fly down lower to give them shots at the target.

Down the beach, the gunners had received instruction to take enough lead on the target before they fired to compensate for the lack of maneuverability of the large and lumbering gunbarrels, pointing fifteen to twenty feet into the air. Unlike the guns in the ball turrets of B–17s which tracked attacking fighters, antiaircraft guns swiveled clumsily in their casing. Often, by the time a gunner could get a target in sight and shoot, the plane had flown on ahead of his line of fire. As Marion began her flight along the beach high above the gun emplacements, the officers walked up and down behind the gunners and bellowed, "More lead! More lead!"

Marion began to hear thudding explosions that seemed to reverberate in her chest. She felt a thrill, of both excitement and alarm, as she clasped the stick of the A–24 and kept the plane at altitude. "So this is what it's like to fly in combat," she thought. Then the Dauntless started to bump in the air, and the explosions became louder. As she looked out the windscreen she saw black tufts dangling in the air around the plane. An instant later, she heard her cable operator shout from the back seat. "Jesus, that's flak! Tell those s.o.b.'s to take a lead on the *sleeve*, not the *plane!*" Marion radioed the gunnery officers that there was flak awfully near her engine cowling. In thirty seconds, the bumping of the airplane stopped. Marion then realized what she had just flown through. The Anti-Aircraft Command motto might be "Keep 'em Falling," but she wondered whose side they thought she was on. Or just how good these gunners really were.

Towing targets, for all the WASPs, was viewed with confusion and ambivalence. None of them had any doubts that flying was dangerous, and that when they joined the Army they were taking risks that most women were not willing to take. Most of them firmly believed that the risks they took were assumed gladly, so that a man pilot could be where he should be, fighting the war. But the gunners below them day after day along Wrightsville Beach, North Carolina, appeared to present, though unintentionally, as much danger as the enemy, which was shooting at their male counterparts for real. The WASPs felt like clay pigeons at a county fair shooting gallery, or more ominously, as they carefully maintained their designated altitudes and courses, like sitting ducks. Among themselves, they tried to make as light as possible of a worsening awareness—complaint would serve no purpose. If one WASP thought her nerve was weakening, she did not want to jeopardize the others' by talking about it. Thus, when a few of the WASPs began losing their appetites, they merely blamed the heat.

Pilots at Camp Davis flew certain antiaircraft training missions at night, to train searchlight operators and radar trackers who would be trying to locate and shoot down aircraft on night bombing raids.

After women had been flying day tow-target missions with no complaints from the antiaircraft boys on the beach, the commanding officer ordered that they be checked out in A–24s at night.

The WASPs had flown many a night over Texas, as part of their Houston and Sweetwater training. But night flying at Camp Davis was different. Since it was along the Atlantic coast, the field had to be blacked out after sunset. There could not even be runway lights. The night traffic pattern had to be very low so that fliers could see the limits of the landing strip. But such a pattern was precarious because of the lofty sea pines surrounding the field.

Some of the women were apprehensive that their airplanes were not in good enough condition for night flying. Like many experienced pilots, some were reluctant ever to fly at night in an airplane that had only one engine. Forced landings were many times more dangerous than during the day when possible emergency landing spots and indications of wind direction were visible. Around Camp Davis, a forced landing at night would be particularly unpleasant. Going down into the swamps, amongst the snakes, bugs and larger creatures of the murk, did not appeal to any pilot at Camp Davis.

For this check ride, the instructors would sit in the gunner's seat and merely observe while the WASPs flew. Each WASP and her instructor perused their Form One sheets carefully. Only minor things were wrong with the planes: a couple of staticky radios, a canopy latch that would not unhook from the inside, broken seats, but nothing peculiar with any of the engines.

Finally, the WASPs and their instructors took off. Marion Hanrahan followed Mabel Rawlinson, a former librarian and singer from Kalamazoo, and her instructor. They all flew around the traffic pattern, practiced a landing, and took off again to try another on the darkened runway. As Marion was entering her downwind leg, she saw Mabel up ahead, flying in the same direction. Suddenly, Mabel's A–24 shuddered in midair. Its landing gear hit the tops of some sea pines. Then the plane plummeted straight down and crashed into the swamp at the edge of the field. On impact, the A–24 cracked into two halves, front and back. The front half, with the engine, burst into flames. Marion, helplessly moving forward, was by this time flying almost directly over the wreckage, eerily ablaze in the dark. Above the roar of her engine, she heard Mabel screaming.

For Marion, eons seemed to pass before she could turn on her base leg, on final approach, and then land her airplane. One by one the other WASPs and their instructor pilots landed and gathered in a huddle in flight operations. The next hour was a blur of ambulances, fire trucks and fragmentary news. When the A–24 had cracked in two, the instructor had been thrown clear of the burning foresection. Mabel, however, encased under her canopy, was charred beyond recognition before the fire equipment could reach her. Her A–24 had been the plane with the faulty canopy latch.

The next day, Tuesday, August 24, Jacqueline Cochran and her executive assistant for the WASP program, Ethel Sheehy, flew to Camp Davis. They spent all afternoon with the commanding officer while the stunned WASPs waited in their barracks for a meeting with Cochran planned for seven o'clock that evening. Marion Hanrahan and Kay Menges were fuming, as were two other WASPs, Mary Lee Logan Leatherbee, an actress and stunt woman, who was the sister of Broadway producer Joshua Logan, and Marie Shale, a former Reno blackjack dealer. All four were savvy professionals who knew a shabby operation when they saw one, and were sharply disappointed in the way their "top-secret experiment" was going so far. They wanted Cochran to hear exactly how upset they were.

After dinner, the WASPs met in a room in the operations building. Some sat tensely in chairs, several paced around the room. No one had slept since the fearsome night before. When Cochran entered the room, she could sense her twenty-five best (now twenty-four) were about to blow. She knew that her plan to broaden flying assignments for the WASPs was at stake.

Immediately someone brought up the deplorable condition of the airplanes at Camp Davis. The very day of the accident, two engines had quit in midair, Cochran was told. The day before, the newest A–24 to the field had an engine failure. Since they had arrived, eleven pilots had suffered forced landings, and the WASPs had spent endless hours flying over the swamps looking for the downed planes. The tires were so old that in one day there were five blowouts. Few of the radios worked.

Mechanics, furthermore, had told them that in their opinions no more than three of the airplanes were airworthy. That kind of talk did not inspire confidence. It was useless, the mechanics told them, to make notations on Form One because they did not do anything about the repairs. They could not get parts. "We are appalled," said one WASP. "We literally fear for our lives." Marion, Kay, Mary Lee and Marie had been speaking alternately, but there were nods of agreement around the room.

Then the subject of instructors was broached. They got the feeling that the check pilots had not had two hours in the ships before taking the WASPs on check rides. Their faith in instruction in general was low, very low. "The reason these men are on tow-target is because they washed out in their course, or had an accident somewhere," someone said.

"Who agrees?" asked Cochran. A shock of hands was raised.

Flight assignments were completely haphazard, the WASPs continued. Though Major Stevenson had finally provided a blackboard to be used by operations officers so that the schedule would be evened up, they paid no attention to it. The whole base seemed to be run by favors. One WASP said that her dentist on the base asked her to arrange a flight for him to another city.

After everyone let off some steam, Cochran was relieved to see the room quieting down. "I'm going to check on those airplanes myself," Cochran assured them. "After the funeral."

At 6:40 the next morning, the women and some of the men gathered in the base chapel for a memorial service for Mabel Rawlinson. Then Cochran met Major Stevenson, a member of the Air Safety Board and the chief maintenance officer to inspect the engine logs of every A–24 on the flightline. They noted the age of each ship and how many hours it had been flown since its last overhaul. Over half of the airplanes were many hours beyond safe maintenance condition for ships used in combat. Mabel Rawlinson's A–24, Cochran learned, had almost 500 hours on the engine since its last overhaul, the equivalent of over 200 target-towing missions.

Another WASP meeting was called for immediately after lunch. When the WASPs filed in, however, Cochran and Sheehy were not alone. Lined up facing them was the base commanding officer, the chief surgeon, and the public relations officer as well. Cochran read figures from her morning's inspection of the flightline, which indicated that the airplanes were not as bad as everyone had exclaimed the evening before. Airplanes stateside simply were not maintained in combat-ready condition, she explained. Most of the WASPs, subdued by the line-up of brass before them, sat silently. Finally Mary Lee Leatherbee spoke up. "I'm not satisfied," she said. "Those engine logs are never up-to-date." The others stared at her with a mixture of disbelief, horror and admiration.

As if there had been no interruption, the base commanding officer began to address the group. He was delighted to have the WASPs at Camp Davis, he told them. He would do everything possible for their comfort and to assist in carrying out their training. The chief surgeon gave the WASPs permission to use the nurses' quarters that they already occupied. Finally the public relations officer requested that he be notified when he could release some publicity about the WASPs at Camp Davis. Then the meeting was adjourned.

Walking away from the operations building, Mary Lee was still angry. "We've been conned," she said. The others shrugged. Lieutenant Alfred Fleishman, in his welcoming speech when they arrived in Houston many months before, had been right. They had to take whatever the Army dished out.

When Cochran was about to board her AT-17 at four o'clock that afternoon to return to Washington, Major Stevenson came running up to the plane. There was some confusion, he said, about flight assignments. According to one of the WASPs, Cochran had told them to fly no equipment on the field until they heard from her further. She assured him no such statement had been made. The WASPs were to fly on orders as usual, unless a particular plane was unsafe to fly as shown by Form One. Then Cochran took off for Bolling Field.

Mabel Rawlinson's accident appeared to confirm all of Major Stevenson's fears about women pilots. For several days, all the WASPs were grounded. Then, tentatively, they were allowed back in the small L4s and L5s, to fly gunnery tracking missions.

Marion Hanrahan decided she had had enough. She had loved flying and airplanes for too long to sour a lifelong enthusiasm. Being a clay pigeon at Camp Davis and forced to fly airplanes that barely merited the definition, Marion realized, was not the way she had wanted to serve her country. The particular character of high command, both at Camp Davis and in Washington, pleased her no more than the A-24s she was flying. She sat down and wrote Jacqueline Cochran a letter of resignation. Several days later, she got a reply. Jacqueline Cochran was not accepting her resignation. Marion had not known that any resignation from such a precarious special assignment so hard-fought by Cochran would be ruinous to her efforts to expand operations of the WASPs. Then Marion and Marie Shale, who joined Marion in protest, requested of the Air Transport Command ferrying division from which they had been released for temporary duty at Camp Davis, that they be allowed to go on duty as ferry pilots after all. This request was also blocked by Jacqueline Cochran, though the ferrying division, knowing of the caliber of the two women pilots plucked out of their hands, very much wanted to have them back. Finally, Cochran arranged for Marion and Marie to resign, but under less than clear administrative skies.

(Marion Hanrahan immediately joined the WAVES. Sent to Honolulu, she worked in the control tower, taught Navy cadets as a

Link instructor, and flew as crew in DC-5s transporting admirals and U.S.O. celebrities to the liberated islands of the Pacific.)

On August 30, five days after Mabel Rawlinson's funeral, twenty-three members of the class of 43-4, who had graduated from Avenger Field on August 7, boarded a train in Washington, D.C. bound for North Carolina. For two days they had stayed in barracks near the Pentagon and been told with great enthusiasm at meetings with Jacqueline Cochran that they would be participating in a pioneering experiment for women pilots. They would be flying missions to train antiaircraft artillery gunners. Girls from 43-3 had already started and were doing beautifully, Cochran told them. But she did not mention any of the problems.

Not until they boarded the train did they know where they would be stationed. Their tickets read Camp Davis, Wrightsville Beach, North Carolina. As they sped through Virginia, before retiring to their berths, they enjoyed a few nightcaps and talked about what they might be getting into. Then they adjourned to their Pullmans and fell asleep.

At two in the morning, Lydia Lindner was jarred awake as her feet jammed painfully against the compartment wall. The car rocked violently. Lights went on and soon the hallways of the cars were filled with people, sleepy-eyed and worried. Another train had rammed theirs from the rear. The last few cars had accordion-pleated, and the passengers who were sitting up had suffered whiplash and other injuries. The young women pilots, physically relaxed by their nightcaps, were unharmed. But the accident did not bode well for the mysterious experiment they were about to join.

The new group arrived at Wrightsville Beach, many hours behind schedule, on the afternoon of August 31. As soon as they were situated in the nurses' barracks alongside the runway, they were called to the operations building. Compared to the spare indoctrination received by the first WASP contingent to Camp Davis, they were briefed interminably by the squadron commanding officer. The newly arrived WASPs sensed that they were being eyed suspiciously. But no one told them of the recent fatal crash, or what to expect from the airplanes they would have to fly.

Soon they were in the air flying L4s and L5s with their predecessors. For ten days, they flew from one end of a small-arms gunnery range to the other, as gunners practiced tracking their planes. Eager to do more, they asked the C.O. if they could attend ground school classes during the hours they were not flying. The C.O., along with many of the men pilots, interpreted this request as an admission that their preparation at Cochran's Convent had been inadequate for the missions they were flying. Nevertheless, classes were quickly organized and access to the Link trainers given for an hour a week. In actuality, the WASPS, who were accustomed to the rigorous schedule of Avenger Field, preferred to be studying for their instrument rating and keeping current on cross-country flying than sitting bored on the flightline. It was beyond the comprehension of high command that, unlike the men pilots suffering the martyrdom of their noncombat assignment, the women were ambitious enough to make the most of their opportunity to fly for the AAF.

In mid-September, the fresh squadron of WASPs finally joined the others in the cockpits of A–24s, and were ready to fly tow-target missions. They soon found that their adventure had only just begun.

As Eileen Roach was part way into her tow-target pattern at one end of the beach at about 1,500 feet, she began to see white specks zing past the nose of her plane and over the canopy. After several rounds she realized they were tracer bullets, nonlive shells shot amidst the live ammunition to show gunners where they were aiming. Then she heard clacking sounds against the metal fuselage. She quickly got on the radio and said, "Hey, I think you are hitting me!" After several resounding metallic whacks, the bombardment stopped. A gunnery officer's voice crackled over the radio. "Sure am sorry," he said affably. "Some gunner down here thought that he was supposed to shoot at the plane up there marked with that cloth marker. We got him straightened out." Eileen flew a few more passes of the beach and then turned back toward Camp Davis. When she climbed down out of the cockpit, she noticed four bullet holes in the fuselage within inches of where her head had been.

After a few days of tow-target missions, the WASPs had landed with shell holes in the tail sections, fuselage, and even in the engine cowling of their airplanes. One WASP returned to base after only one pass of the beach. She was rather shaken. A direct hit by a .90-millimeter shell had exploded her target completely off its cable. She had watched it fall blazing into the ocean as she envisioned her plane blown to pieces. Though at first they had been disappointed that they would not be ferrying airplanes for the AAF, the target-towing WASPs had now decided that their special assignment was far more exciting.

Even when they began to learn first-hand about the hazards of flying at Camp Davis, they thought they should act like the coolest pilots in the Army. One day after a tow-target mission, Lydia Lindner landed her dive bomber tow ship. She had found that landing hot, or at about 110 miles an hour, she felt safer. On slower approaches, the wind buffeted against her wings as if they would stall. As she screeched onto the runway, however, she was suddenly jerked to one side, as if she were in a sharp turn on the Coney Island roller coaster. Lydia braced herself with her hand against the instrument panel, cut the engine and played frantically with the rudder pedals as the plane careened off the runway. Finally it jolted to a stop on the grass. She threw open the canopy and jumped down to the ground to see what had happened. Checking her undercarriage, she saw immediately that she had blown a tire. Firetrucks and a jeep driven by the officer of the day raced toward her. "Are you all right?" the O.D. shouted when he got within earshot. "Oh, sure," Lydia yelled back, "I'm fine." When the O.D. reached her, he asked again. "Yeah? Are you sure you're all right?" Lydia reiterated that she was just fine. The O.D. was now grinning. "Do you always smoke two at a time?" he asked. Lydia looked down to see two lit cigarettes in the same trembling hand.

In spite of the ribbing she took for the next few days, Lydia soon learned that when it came to gumption under fire, she was getting it, especially when challenged as a woman pilot. Most of the servicemen who operated the target reels had made flights with WASPs by that time, but when the twenty-two-year-old blue-eyed former Brooklyn College coed sauntered out to her tow-target ship one afternoon, she noticed that her reel man's eyes widened in apprehension. When she tried to make friendly conversation as they prepared for takeoff, the young GI answered her only in monosyllables.

Out over the beachhead, Lydia flew a wide circle so that the cable operator could reel out the target sleeve before they started their pattern past the gunnery positions. Then they headed down the beach. Just as the guns began to fire, Lydia smelled smoke. Her eyes darted around the instrument panel and underneath by the rudder pedals, but she saw nothing extraordinary. Then she heard a strangled-sounding voice from the back seat. "There's a fire back here, don't panic, we can probably get in." Lydia turned around as casually as she could. Smoke was coming out from under the gunner's seat near the cable winch. Sparks from reeling out the cable must have set fire to the seat. From one glance at the reel man's pale young face, she could read his mind. It said, "I'm up here with a girl pilot for the first time. I've never met her before. I'm scared to death of what she is going to do in a panic situation. She's probably going to jump out and leave me here. . . ." At 1,000 feet, neither one of them had altitude enough to jump. They had to make it back to the base. "Roger," said Lydia in her most businesslike tone, and called the tower. "I have a little fire," she said as if asking for the time. "Okay," said the tower. "Cleared for an emergency landing." They cracked the canopy as the smoke billowed around the cockpit. Lydia could hear the cable operator thumping energetically in the rear cockpit. She concentrated solely on making the softest landing she could. With two gas tanks in the center section of the airplane, she did not want to risk an explosion. If another tire blew, she and her reel man would be statistics, under nonbattle deaths. Thinking of her young passenger, however, Lydia suddenly felt very brave. She made the gentlest of landings, and by the time the fire trucks reached the A–25, the cable operator had the fire under control. He hopped down out of the cockpit, trembling, and obviously a relieved young man. As for Lydia, she climbed down, put her hand on his shoulder, and looked into his eyes. "You said we could make it, and we did." He smiled weakly, and Lydia walked calmly into the operations office.

One afternoon Kay Menges was given an assignment to fly an A–24 one hundred fifty miles north to Rocky Mount, North Carolina, on a delivery for one of the officers at Camp Davis. She did not mind being a messenger; she welcomed her first breather from Camp Davis since her arrival three months before. The early fall colors promised to be lovely, but she also looked forward to a day in the air to reflect on all that had happened over the past several weeks. When the second group of twenty-five WASPs began flying tow-target missions, Kay Menges was relieved. There must be strength in numbers, she thought. The tenor of the flightline was much more relaxed since 43–4's arrival.

When she returned at sundown, however, she found Camp Davis again in an uproar. In the operations building everyone was running to and fro and wearing worried looks. There had been another crash; another woman pilot had been killed. Kay met one of the WASPs, Helen Snapp, who had just come in from her mission flying the beachhead. Helen's large wide-set eyes were troubled. While she was circling the field to land, she said, she saw an A–24 on its back in the grass off the runway. There had been no fire, miraculously, but Betty Taylor, a WASP from California, and her passenger, the Camp Davis chaplain, had been crushed when the weight of the plane smashed the canopy.

Betty's death had evidently affected Helen deeply. In between their graduation from Sweetwater and their special assignment to Camp Davis, Betty had been married. Now, a month later, she was dead. Helen, too, had recently married. When her husband, Ben, went overseas with the infantry, Helen had joined Jacqueline Cochran's pilot training program. Since then, Ben had been among the first American troops to land in North Africa and, she had just learned, had been with the American force that had invaded Italy. The armistice in Italy had been announced a month ago. Helen had disciplined herself firmly not to think about the possibility of Ben's being killed in action. But she was thunderstruck at the idea that she might not survive the war.

Kay Menges decided she would pay a visit to her mechanic friends to see what had happened with Betty Taylor's airplane. She knew that the A–24 was a tricky plane, and that Betty had only recently had the WASPs typically perfunctory check-out in it. The engine had to be well warmed up before taking off or it might quit. Also, if a landing had to be aborted, the pilot could not count on giving the plane full power in order to go around the traffic pattern for another try; after being throttled back, the engine often did not take when gunned. Kay suddenly remembered that she had recently flown an A–24 that had a sticky throttle. Slow enough normally when pushed forward to full power, this plane's engine did not respond for several seconds after she hit the throttle. Then all of a sudden, the plane roared forward and lurched into a climb. She had to jump with all her weight on the right rudder or the sudden torque of the propeller would have spun the plane right over.

From the mechanics, Kay learned that, on final approach, Betty had suddenly announced a go-around. But instead of roaring into a climb, the plane had hit the runway hard on the main gear, bounced, and flipped over, to come crashing to the ground on its canopy top. Kay asked to see its Form One. There, in her own handwriting, were the words, "Sticky Throttle." Not surprisingly, no repairs had been made. But Betty had not known how to correct for the delayed surge of power when she gunned the engine for her go-around.

The next day, Jacqueline Cochran again flew to Camp Davis from Washington. With her two most vehement opponents resigned, she did not meet with the same intense, near mutiny from her WASP squadron. Instead, she saw merely sadness about the death of the quiet, shy young Californian who had run out of luck. By now the fact of death was sinking into the women's emotions. Many from 43–4 had known Jane Champlin, the St. Louis woman who had died with her instructor on a night flight while training in Sweetwater. The others had experienced Mabel Rawlinson's crash merely a month before. They were beginning to learn that a WASP had to protect herself. She had to know the peculiarities of a red-lined airplane, not fly when she did not trust one, and know how to get herself out of trouble.

Cochran, nevertheless, investigated Betty Taylor's accident. After the funeral the next afternoon, she left Camp Davis without a word about her findings. What she had uncovered could not be discussed. And she had asked mechanics with whom she had made the discovery to keep quiet and not to alarm the girls. If they found out, Cochran feared, there would be an insurrection far worse than that after Mabel Rawlinson's death, and the entire WASP program might be thrown in jeopardy. In the gas tank of the demolished A–24 mechanics had found traces of sugar. Sugar in sufficient quantities could stop an engine in seconds; even a teaspoon could cause a rough response. The "sticky throttle" Kay Menges warned of now had ominous implications. Cochran was convinced the accident was caused by sabotage. But she did not know against what or whom the effort had been aimed. And she never wanted to find out.

A second fatal accident of a woman pilot inflicted a permanent blow to the commanding officer's already shaky confidence in the WASPs stationed at Camp Davis. Though they would remain in Wrightsville Beach until May 1944, none would be allowed more than copilot duties on any of the larger ships on the field, the twin-engine C–45 transport and the B–34 medium-range bomber. Word that WASPs were successfully ferrying these planes with the Air Transport Command, if it made its way through the swamps to Camp Davis, did not change the situation. Also on the flightline, and gazed on dreamily by the WASPs, were P–47 Thunderbolts, the AAF's heaviest fighters, used for strafing missions over gunnery implacements. Though WASPs in the Air Transport Command had been ferrying them since the summer of 1943, and dozens were about to enter pursuit school, Major Stevenson would not let his WASPs near his P–47s.

Though high command remained intransigent, the other men on the field began to experience a change in attitude. As the WASPs jumped up to volunteer every time the operations officer offered a flying mission, the men pilots, still disgruntled at their banishment, found they could now sit and continue their card games. Some days, the eager-beaver WASPs flew both a morning and an afternoon or a night mission—adding up to as many as five or six hours in the air a day. When gunnery classes were under pressure, pilots were on call to fly weekends. WASPs took those missions, too.

For the men pilots, an insult had quickly turned into a blessing. No less appalled than those WASPs who had raised a furor with Jacqueline Cochran about the flying conditions at Camp Davis, the male officers quickly deferred. Understandably not wanting to take any more risks than necessary in this noncombat assignment, they saw the advent of fifty (now fewer) capable WASPs as the answer to their unspoken prayers.

The enlisted men who in July had swarmed the commanding officer and demanded transfers had also experienced a change of heart. Though Major Stevenson kept his promise to arrange for their transfers, one young cable operator replied for the rest. "Well, sir," he said, "we think maybe we better stick around here and see that these girls get through this damned course."

And there were very few complaints from the beachhead. Artillery officers began specifically to request WASPs for all kinds of tracking missions. Planes instructed to fly figure-eight patterns over the gunnery range were routinely disappearing from sight and young gunners were flunking their exams. The male pilots, bored with the job, would break out of the pattern and fly off to practice acrobatics. Women pilots, however, would perform figure-eights, dizzying as they were, with precision until dismissed. So WASPs were wanted on those missions.

One afternoon, WASP Lydia Lindner assumed supremacy in the most spectacular tracking mission of all, that of simulating dive-bomber raids on the antiaircraft artillery gunners.

After many hours in the A–24 and weeks of persistent requests, WASPs were allowed to fly the A–25 Curtiss Helldiver, a 1,700-horsepower dive bomber used by the Navy. It was built like the Dauntless, but it was far more powerful. The A–25 was known informally in the pilot's ready room as the "coffin." For weeks, WASPs had heard men pilots joking that they had to "go up in the 'coffin' today." Diving missions in the "coffins," many of which were red-lined, were not prized.

That afternoon, the pilot assigned to do the day's diving mission over gunnery positions on the beach did not appear at his appointed time. The operations officer looked nervously at his watch. Having lined up the gunners along the beach only to have the mission canceled innumerable times, gunnery officers were suspicious of their pilots, whom they felt were lazy. They had their quota of trained gunners to fill, and the pilots did not give a damn. The operations officer who assigned pilots often took the flak.

Lydia Lindner had seen A–25s dive at gun positions, pull up and soar off to come around and dive again. Cameras in the barrels of the guns along the beach, she understood, recorded how well the gunner tracked the attack. Dive missions looked exciting to her. She jumped up and offered to take the male pilot's place. The operations officer looked more worried than ever. "I don't know about that," he said, frowning. "Have you ever done a dive mission?" "No, what's so hard about it?" Lydia asked. The operations officer shrugged. "Nothing. If you want to do it, go ahead."

Lydia found an A–25 that was in good condition, hopped in and adjusted her throat microphone on its collar. She knew the gunnery officers might send her back if they knew a girl was doing a dive mission. Having trained for the stage and played Shakespeare in high school in Brooklyn, Lydia cleared her throat and prepared to speak in low, modulated tones. As she approached the beach, she reported to the artillery officers, with all the stage presence she could muster, that their new dive mission pilot had arrived. They told the unfamiliar Helldiver pilot to climb to 8,000 feet and dive down to about 200 feet above the beach. "Just make believe you are strafing us," the gunnery officer said.

Lydia circled and climbed. As she got to altitude, she pushed the nose of the A–25 Helldiver into a dive. She had never felt such an oneiric sensation. She watched the beach come closer and closer. Her whole body felt electrified, as the engine roared around her. She imagined herself a fighter pilot diving onto a Japanese-held island.

Suddenly, as if coming out of a trance, she realized she had to pull back on the stick. As she pulled, however, the seven-ton Helldiver kept diving. Lydia braced herself, her feet against the floor, and pulled against the stick with all her might. The shallow breakers were still rushing toward her. "I'm done! I'm finished!" she gasped as she began to feel the G-load increase on her body. Then she blacked out. Several seconds later, as she came to consciousness, she was careening level with the Atlantic and beginning slowly to climb again. Her altimeter read fifty feet. The stick was still pressed into her stomach, and she felt as if she weighed 500 pounds.

A breathless male voice cracked in her earphones. "Oh, that was great! Oh, man! Do that again!"

Lydia began to feel embarrassed by her brush with death. As she dove again at the beachhead, she kept her exuberance in check and started to pull the Helldiver out of its dive a little sooner. By the end of the mission, she had the maneuver down so precisely that she felt qualified for combat. It had been some day.

When she reported to flight operations the next morning, the officer grinned at her. For the dive mission, he said, the artillery boys asked specifically for the pilot of the day before. He chuckled. Now they were bound to find out who their pilot was. Indeed, Lydia's secret was not kept for long. At the end of that afternoon's dives, she heard in her earphones, "Hey, WASP, you're doing a great job. It's more cooperation than we've gotten from those guys they've been sending us." Lydia knew what they really meant: she was the first A–25 dive bomber pilot they had been able to find who was crazy enough to simulate an actual attack. The men pilots were cautious. Nevertheless, Lydia found herself beaming, all by herself in the cockpit of her "coffin."

Three months after the first women pilots had arrived at Camp Davis, Jacqueline Cochran decided to release her experiment to the press. In late October, accompanied by several AAF generals from the Pentagon and First and Third Air Force headquarters, Cochran flew into the Carolina swamps to review the WASP tow-target squadron. The national press was invited to spend the day on the field.

When they learned of this onslaught of gold braid, Major Stevenson and his public relations officer decided that the WASPs should have their own recreation room in which they could relax on the flightline between flying missions. The WASPs flying tow-target and tracking missions that afternoon returned to find that a storeroom above the flightline ready room had been cleaned out and furnished with some card tables and chairs. On the door was a sign which read, "WASP Nest—Drones Keep Out or Suffer the Wrath of the Queen."

On Thursday, October 28, 1943, the forty-six WASPs of Camp Davis stood at attention in front of two huge B–34 bombers. The Director of Women Pilots and the generals walked casually along the columns of pilots who were wearing improvised uniforms of gabardine slacks, white shirts and flying jackets. The brass smiled warmly at the WASPs and leaned over toward Cochran to ask questions as they passed. They were evidently impressed. Following Cochran and the brass were Major Stevenson and members of his staff. As they reached the last WASP in the line-up, the major slapped her on the buttocks and hissed playfully, "Get!" The WASPs dispersed, followed by several reporters and photographers.

On October 29, the WASPs of Camp Davis were described in an

article by an Associated Press reporter. Eloquently, he told of a WASP climbing into the cockpit of an A-24 Douglas Dauntless: "She held her head high. Her face was tanned by sun and weather. The wind blew her dark hair straight back. She had the air of serene confidence that distinguishes fliers." And "pretty Eileen Roach of Phoenix" was quoted as saying, "We get a thrill every time we leave the ground."

Ebullient on America's valiant women fliers, articles continued appearing for the next few months in papers across the country. But the day after Jacqueline Cochran, the generals, the reporters and photographers departed from Camp Davis, the sign was gone from the door of the "WASP Nest." Their lounge was once again a storeroom, and the WASPs were back in the air, towing targets over Wrightsville Beach.

XIII

Cool Pilots for Hot Ships: The Most Dangerous Game of All

The week before Christmas, 1943, in Palm Springs, California, Nancy Batson climbed into the single seat of her first P–47 Thunderbolt. Her instructor, squatting on the wing beside her, gave her a last-minute cockpit check. Nancy lifted a trembling hand to point to an unfamiliar knob, then quickly pulled it back into her lap before her instructor noticed. She felt as if she were sitting in a barrel about to go over Niagara Falls. One of the original WAFs at New Castle, Nancy Batson had been flying Army airplanes for the AAF Air Transport Command, first trainer aircraft and then twin-engine cargo transports, for fourteen months. But she had scarcely dared dream she would become a pursuit pilot.

The P–47 was the biggest, heaviest single-engine pursuit-type, or fighter, airplane in the Allies' flying armada. Weighing 12,500 pounds, the Thunderbolt was twice the size of the famous British Spitfire, and its 2,400-horsepower engine drove a thirteen-foot, four-bladed prop which loomed blindingly in front of Nancy's eyes. Because of the Thunderbolt's thick rounded form, it was affectionately called the "Jug" by fighter pilots, many of whom were already attaining Ace status as they flew along as protection for huge lumbering B–17s and fought off German interceptor fighters, like the Focke-Wulf and Messerschmitt, in deadly aerial skirmishes over Nazi territory.

On December 1, 1943, Nancy Batson and nine other WASP pilots were picked by Nancy Love, director of women pilots in the Air Transport Command ferrying division, to be the first group of women to attend the ATC pursuit school. For three weeks, Nancy had prepared herself for the day she would take off in her first fighter. As the P–47's snug cockpit provided room for only one, her first flight would be solo. Her training, simulating pursuit conditions as closely as possible, had been while flying an AT–6 from the back seat. She had learned to taxi with the AT–6's nose thrust high in front blocking her view, forcing her to S-curve down the taxiway, to see where she was going. To approximate a pursuit-type landing, she was taught to dive the AT–6 at 120 miles an hour straight toward the runway and pull back on the stick at the precise moment when the wheels would meet the concrete. Experts from Pratt-Whitney and Republic Aviation had grilled her on the power plant and flight characteristics of the fighter. But now she was sitting in a real P–47 Thunderbolt and she had to handle it alone.

Her instructor hopped off the wing, and a mechanic stuck a battery charger into the side of the engine. Nancy pressed the starter button, pushed the throttle forward, and as flames burst out of the exhaust pipes, the giant engine roared to a start. "Oh, God," Nancy prayed, "I've got to go now!" Pushing the button on the stick to activate the throat mike, she called the tower and announced that she was ready to taxi.

Craning her neck to the side, Nancy maneuvered the throttle and, working the rudder pedals, slowly began to S-curve the six-ton fighter down the taxiway. She felt as if she were looking from a second-story window. The P–47's tail wheel swiveled, making taxiing easier than in the AT–6. But this fleeting observation did nothing to calm Nancy's nerves.

Reaching the takeoff end of the runway, she went down the pre-takeoff check list carefully. When she ran up the engine to check the magnetos to the spark plugs, the manifold pressure and rpms, the noise in the cockpit was so deafening, she thought she might faint. All too soon, her check list was completed, and she had nothing else to do. She sat for a full minute listening to her heart pound above the din of the engine. "Now, I really have to go!" she gasped. She called the tower and was cleared for takeoff.

Nancy took hold of the stick, settled her feet on the rudder pedals and took a deep breath. She could see nothing beyond her instrument panel but the huge engine. She pushed the throttle all the way forward, and the P–47 began to surge down the runway. As the airspeed indicator climbed, the noise shook her whole body and she was pressed back against her seat. Then she lost consciousness.

When Nancy came to herself, she was at 10,000 feet over the Salton Sea. She pulled the throttle back gingerly, trimmed the pressure off the flight controls and looked into the flawless blue desert sky all around her. She had the sensation of floating, though her airspeed indicator told her she was traveling at almost 300 miles an hour.

"Well, here I am up in this thing," Nancy said to herself." Let's see, I'm going to make just a little turn . . ." She moved the stick delicately and the P–47 banked to the right for a few seconds as the desert rose over her right shoulder. Then as she moved the stick slightly to the left, she recovered from the turn and was again speeding along over the desert. For several seconds she sat wide-eyed. With another burst of determination, she decided to make another little turn. Again the fighter banked and righted itself like drifting in a dream.

"Oh, heck!" Nancy shouted. "This is just an ol' airplane!"

Nancy Batson, overcome with awe as she took off in her first pursuit, shared the feelings of every new fighter pilot during World War II. The heroics of pursuit planes were renowned from the early moments of the war. The Battle of Britain had been won by the scrappy Spitfire and Hurricane fighters who, with each wave of German raids, took off in pursuit of the JU 88 bombers before they could reach London. The P–40 Warhawks of General Claire Chennault's American Volunteer Force, known popularly as the "Flying Tigers," were legends both for the engine cowlings, which were painted like the gaping jaws of a shark, and for their fearless forays against the Japanese over the Hump in China.

The men who flew pursuits in AAF Fighter Groups were revered as the proud and flamboyant heirs of World War I Aces, barnstormers and the speed demons of the 1930s national air races. Fighter pilots had to perform the most dazzling aerial acrobatics—*and* shoot with deadly accuracy at the same time. Bomber pilots had to be blessed with stamina, fortitude and courage for the long run, and the leadership quality to inspire their crews to bravery. Pursuit pilots flew, fought and died alone. Their airplanes were built for speed, spurts of high performance and maneuverability; their missions were to peel off and dive through the flak-riddled sky to strafe bridges, railroads or ships, and to dogfight, as fearless, agile escorts to vast armadas of up to a thousand bombers on their way to their targets deep behind enemy lines. An air corps aphorism joked that if a man could count to ten, he could not be a pursuit pilot. He thought too hard.

World War II's pursuit planes developed from the one-of-a-kind, record-breaking racing planes of the 1930s, like the P–35 which brought Jacqueline Cochran victory in the 1938 Bendix Race and then became the P–47. Fighters were all engine and firepower. Not built for economy, pursuits' high-performance engines, like the P–47's, burned sixty gallons of hundred-octane fuel an hour. Early in the war, their range was limited to a few hundred miles. Because a fighter escort could only be partway, bombing raids had to be conducted at night.

But in the summer of 1944, P–47s began to roll off the assembly line with droppable auxiliary fuel tanks, giving them a range of 1,100 miles. Immediately, the Eighth Air Force began intensive daylight raids, which were brasher, but far more accurate than at night, on industrial targets deep into Germany. In March 1944, with the AAF's new P–51 Mustang long-range fighter, B–17s and B–24s would be escorted all the way to Berlin, a 1,200-mile round-trip from the British Isles.

The cockpit of a fighter, measuring no more than three-by-three feet, was a confusion of dials, levers and knobs. Every inch of space was used. Even the stick, more than just a lever connected to the ailerons and elevator, was a fist-sized control center, with buttons to activate the throat microphone and to fire the machine guns in each wing. Around the pilot's legs were levers to work the hydraulically controlled landing gear, manual back-up systems, trim tabs on the ailerons, flaps on the elevator, and other controls which released bombs or empty fuel tanks from under the wings or fuselage. The entire cockpit seemed to be plastered with glass-covered gauges, monitoring countless functions of the immense and intricate engine: gasoline consumption, temperature or pressure of engine coolant, hydraulic fluid, oil, gyroscopic vacuum, exhaust gases and superchargers. Then there was all the communications and navigational radio equipment. In fact, there was barely room for the pilot; many fighter pilots were small men, not prone to claustrophobia.

Each pursuit plane was worth around $100,000 (the equivalent of over a million dollars today), or about $2,500 a foot. And in spite of their awesome reputation, advanced technology, precision flight characteristics and often temperamental, high-performance engines, thousands of pursuit planes, fresh off assembly lines, were flown on their first legs toward combat zones in Europe and the Pacific by women. After trainer aircraft, World War II's WASPs flew more P–39, P–40, P–47, P–51 and P–63 fighters than any other airplanes. But for these unsung fighter pilots, who shot down no Messerschmitts or Zeros, their skirmishes were on the home front.

When women pilots were first allowed into the cockpits of military airplanes in September 1942, they were assigned by the Air Transport Command to ferry small trainers. By December 1943, six months after Jacqueline Cochran was appointed Director of Women Pilots in the AAF and consolidated the WAFs and her training school into the WASP, Avenger Field had poured almost 200 women pilots into the Air Transport Command. Cochran had also fulfilled her ambitions to broaden assignments for her WASPs to include training in the AAF's biggest bombers and missions towing targets for antiaircraft artillery units, one of the toughest operational flying duties in the AAF. But Nancy Love, founder of the WAFS and now commander, under Cochran, of the WASP squadron in the Air Transport Command, had seen many battles of her own in the AAF organization women pilots set out to serve.

As the WAFs gained experience and confidence flying PT–19s and L4 Cubs during the winter of 1942–43, they became aware that the "hot pilots" of the Second Ferrying Group at New Castle, Delaware, were the men checking out in pursuits. For weeks, the WAFs gazed at the huge P–40s and P–47s sitting on the flightline and fantasized about flying them. So did Nancy Love. But when she asked Colonel Robert Baker, commanding officer at New Castle, if she and the WAFs could check out in the P–47, being ferried in increasing numbers from the Republic factory in nearby Long Island, he shook his head. Though the compact cockpits most comfortably accommodated smaller body frames, Air Transport Command headquarters thought that the high-powered pursuit engines could not possibly be controlled by a woman.

When Love established the fourth (and eventually the biggest) female ferrying squadron in Long Beach, California, in February 1943, she marveled at the variety of manufacturers in Southern California whose aircraft were being ferried by Long Beach's Sixth Ferrying Group: Lockheed, Douglas, North American and Consolidated Vultee. Here, far from headquarters politics, Love felt more free to maneuver. On February 27, 1943, as the women of 43–5 were packing their trunks to report to Sweetwater, and long before Jacqueline Cochran was to launch her official WASP experiments, a lone P–51 Mustang fighter took off into the pale Pacific skies, with Nancy Love in the cockpit. Not only was she the first woman, Love was one of the very few American pilots to fly the Mustang in early 1943. First manufactured for the RAF in Britain, the P–51 would not be flown by AAF fighter groups until December. While Nancy Love put her P–51 through its paces that February morning, the Mustang Aces of the future were still in AAF cadet school.

The P–51 was only the beginning. As experienced pilots, the WAFs, naturally, wanted to fly anything with wings, and the official restriction to AAF trainers was beginning to gall them. Within a month of her arrival in Long Beach, Love had checked out in sixteen other types of airplanes, including the C–47 twin-engine cargo transport, the military version of the famous DC–3. Other WAFs followed her into each cockpit. The Sixth Ferrying Group saw women simply as pilots, much needed to fly the vast number of airplanes produced by the manufacturers it served. Love thought that the WAFs' days in the tiny trainers were over.

Even back at New Castle, the WAFs' squadron commander Betty Gillies was allowed to check out in the P–47. Then quietly, and unofficially, WAFs Gertrude Meserve and Teresa James were also given ground instruction by a cooperative check pilot.

Male officers at ferrying division headquarters began to worry about women's physical and emotional capacity to fly valuable AAF fighters and multiengine airplanes. This was when a letter went forth (on March 29) to all ferrying group headquarters which officially prohibited assigning women flying duties between one day before and two days after their menstrual periods—in effect, ordering them grounded for at least a week every month. The Air Surgeon General, who had not been consulted on the prohibition, promptly overruled the ATC and ordered that menstrual periods should be regarded as an individual matter, not a command function.

Male officers at ATC headquarters were not alone in their apprehension about women pilots. On March 25, 1943, the commanding officer of the Third Ferrying Group at Romulus, Michigan, issued a comprehensive memo to his flight operations personnel, setting forth exactly what the five WAFs on his base were and were not allowed to do. First he ordered that women pilots would fly only trainer aircraft. They would be given no transition to pursuit aircraft or twin-engine bombers, and no copilot duties, either. Women and men pilots were to be assigned to flights on alternate days, or if that were impossible, sent in opposite directions. No mixed crew assignments would be tolerated.

Nancy Love, who, as the itinerant WAFS leader, paid regular visits to her four small WAFS squadrons, arrived in Romulus just in time to see the memorandum. She was furious. If the blanket restriction to AAF trainers was not definitive enough, the rest of the memorandum, namely the ban on mixed pilot and crew assignment, virtually guaranteed that women would never fly airplanes larger than trainers. Even if her experienced WAFs could train each other in twin-engine planes, progress would be absurdly slow. And on larger bombers, there was usually a flight engineer or navigator, who, of course, was male.

In spite of her personal efforts to prove women's flying ability, the ferrying division seemed to be clamping down more than ever. Love decided to appeal directly to the Air Transport Command Chief of Staff in Washington (who was, incidentally, her husband's immediate superior).

Two weeks later, on April 17, General William Tunner, head of the ferrying division who, with Love, had formed the WAFS, was politely directed by Air Transport Command headquarters:

> It is the desire of this command that all pilots, regardless of sex, be privileged to advance to the extent of their ability in keeping with the progress of aircraft development. Will you please insure that the terms of this policy are carried out insofar as it applies to ferrying of aircraft within the continental U.S.A.?

It is unclear how much General Tunner participated in the discussions among his group commanders and headquarters staff before receiving this order. He may in fact have been waiting for it, as ammunition against mutiny for what he was about to do. On April 26, Tunner issued a comprehensive set of WAFS regulations for all of his ferrying groups to follow. Though restrictions against men and women in the same cockpit remained, he exempted training flights, and demanded that women pilots be transitioned to multiengine and pursuit aircraft "under the same standards of individual experience and ability as apply to any other pilot." Suddenly, women pilots were not only allowed to fly everything in the ferrying division's coffers, they were *ordered* to. Tunner's staff was outraged. "*Even* the P–51?" asked one.

Unlike his underlings who were reacting to women pilots *per se*, Tunner realized he had to have their long-range utilization in the ferrying division in mind. At that time, he was putting the final touches on an ingenious on-the-job training program for ferry pilots. Delivery orders were so urgent, he could not spare much pilot time specifically for training. Thus, he arranged AAF aircraft into Classes I through V, from small trainers to four-engine bombers, and ordered his pilots to spend a certain amount of time ferrying planes in each class. The ultimate proficiency was a Class V—overseas rating. But every new pilot would begin flying Class I and II AAF trainers. During this initial phase, ferry pilots got crucial cross-country navigational experience, while delivering much needed aircraft to AAF training bases, a top priority in the spring of 1943. (General Tunner's on-the-job training program would prove so successful in turning out crack pilots that the AAF Commanding General would pluck out some of

Tunner's experienced multiengine rated pilots for overseas combat duty, to supplement the supply from AAF advanced transition schools, like the B–17 school at Lockbourne Army Air Base.)

But as 1943 progressed, the demands of Allied bombing strategy leading up to an invasion of Europe called for more and more bomber and fighter pilots in the ferrying division, to deliver combat planes overseas or to American ports of embarkation for placement on shipboard. All these men had to go through Classes I and II cross-country orientation. General Tunner began to look with concern at the huge classes of over a hundred women entering Jacqueline Cochran's training program, whom he thought at that time to be headed for the ferrying division. If the women were all frozen as Class I and II pilots, not a seat would be left for his men pilots to begin training. Women had to progress, had to be spread throughout the system. The ferrying division motto read, "A pilot for every plane that's made, over every route that's flown, to every fighting front." For the hard-pressed General Tunner, this motto included his top-flight women pilots.

WITH HER APPOINTMENT as head of WASP ferry pilots in July 1943, Nancy Love moved into her new office at Air Transport Command ferrying division headquarters, now located in Cincinnati, Ohio. At this time, the Eighth Air Force based in Britain requisitioned a hundred B–17 Flying Fortresses, and the Air Transport Command ferrying division began a "blitz" of transatlantic bomber deliveries. Tunner's young male ferry pilots, freshly qualified to fly the four-engine bombers and assigned to the blitz, faced the expanse of the Atlantic with trepidation. Such a flight was still conceived in mythical proportions performed successfully only by singular godlike figures remembered from their childhood. In order to accomplish this prodigious movement of bombers, Tunner pondered a way to raise morale and confidence among his pilots. Suddenly he remembered an incident that recently had occurred at his Third Ferrying Group, Romulus, Michigan.

Ferry pilots at Romulus were responsible for delivering most of the P–39 Airacobra fighters manufactured by the Bell Aviation factory in Niagara Falls, across Lake Erie from Detroit. The two dozen women pilots were assigned to the trainer and liaison aircraft that happened to be at the base, while the men ferry pilots bore the brunt of P–39 deliveries, which were increasing every week as the Bell assembly lines geared up to supply America's Russian allies. But the P–39 was proving to be treacherous. It's 1,200-horsepower engine was mounted behind the pilot, rather than in front as in other pursuits, and directly under the cockpit were large fuel tanks. Thus the weight was further to the rear than the men were used to. An alarming number of Tunner's pilots had not followed the tech orders precisely while taking off or landing. Letting the light nose of the P-39 rise too high, they had spun in, crashing to their deaths. Calling the Airacobra a "flying coffin," pilots were contriving excuses not to fly deliveries of them. Then a WAF stationed at Romulus wangled her way into the cockpit. For a day, she practiced takeoffs and landings, experimenting with various airspeeds, and found that the plane had to be landed at a higher speed than the men had been using. With her successful check-out, she helped two or three other women pilots train in the pursuit. In June, women had delivered three P-39s from the Bell factory to Great Falls, Montana. Since then, miraculously, landing and takeoff accidents in the Airacobra had virtually stopped.

Tunner picked up the telephone and called Nancy Love into his office. "I want you and another WAF to fly a B-17 to England," Tunner told Love. "I've got a morale problem on this blitz to the Eighth Air Force and you're going to help me solve it."

By August 15, Love and WAF Betty Gillies, her squadron commander at New Castle, had joined the few British women ATA pilots as the only female four-engine bomber pilots. (Cochran's B-17 experiment would not begin until October.) The men in the ferrying division transition crew marveled at the skill and stamina of the two women pilots, especially the five-foot-one-and-a-half, 108-pound Gillies who, with cushions behind and under her, could hold the bomber level on a three-engine procedure like any AAF officer twice her size.

On Thursday morning, September 2, Nancy Love and Betty Gillies, with General Tunner's hand-picked flight crew, boarded B-17E Number 42-30624, the Flying Fortress they would deliver to the Eighth Air Force in Britain. Their orders, which were from New Castle to the U.S. Air Transport Command base at Prestwick, Scotland (where two years earlier Jacqueline Cochran had ferried her Lockheed Hudson bomber) had been kept top-secret, so as not to incur press attention. Word of mouth among men ferry pilots, that two "gals"—one of them a diminutive five footer—had flown a Fortress to Britain, would be sufficient to allay the men's fears of the bomber blitz's many transatlantic flights. The two women pilots were thrilled. Among their select crew was General Tunner's personal navigator, Lieutenant P. O. Fraser, a feisty character who insisted the two young women call him "Pappy."

General Tunner took the precaution of secrecy for reasons other than press attention, however. Though he knew that there was no formal legislative or military prohibition against women flying airplanes overseas, he had specially cleared Love and Gillies through the AAF Commanding General's office. General Arnold was in England at the time; in his stead, General Barney Giles, the AAF Chief of Air Staff, had given his approval.

Then, to guarantee a successful flight, General Tunner not only gave Love and Gillies his navigator, he also modified the standard ATC transatlantic ferrying route to give them shorter over-water hops. Rather than flying from Delaware to Presque Isle, Maine, on to Gander, Newfoundland, and then straight to Prestwick, Tunner ordered them to make stops, after Presque Isle, in Goose Bay, Labrador, in Greenland and in Iceland before flying on to Great Britain.

Frustratingly, the trip up the coast from New Castle took three days because of stormy weather. But as Sunday, September 5, dawned in Goose Bay, Love, Gillies and their crew boarded the B–17 and went through a final cockpit check. Suddenly, out the lofty cockpit window, Love saw the Goose Bay base commanding officer running toward the plane and waving frantically for them to get out. On disembarking, Love was handed a radiogram from General Barney Giles, the AAF Chief of Air Staff. On behalf of General Arnold he was canceling their mission to Scotland. They were to hand over B–17 Number 42–30624 to an all-male crew. Two hours later, bewildered and numb with disappointment, Love and Gillies watched their B–17 take off, with two men ferry pilots in the cockpit, for Prestwick.

"Where on earth were you? We looked all over Scotland for you two!" the ferry pilots back in Presque Isle, Maine, shouted when Love and Gillies arrived by transport. Their jovial greeting was cut short when they saw their female colleagues' faces. A gulf of silence hung between the men and women ferry pilots that was far too broad for words to bridge.

In a gallant effort to reduce the hazards of transatlantic flight for his two top women ferry pilots, General Tunner left his mission vulnerable to the most uncanny turn of fate. As soon as his women B–17 pilots were safely off from New Castle the previous Thursday morning, General Tunner had notified ATC headquarters. General George's deputy had in turn wired Brigadier General Paul Burrows, commander of the ATC European Wing, that a B–17 was en route to Prestwick piloted by two WASPs. The message arrived while General Burrows was having dinner. Astonished at the contents, Burrows handed the wire to his dinner companion, his boss, Chief of the

Army Air Forces, General H. H. "Hap" Arnold. That was the end of their cordial dinner. General Arnold shot up from the table and raced to the telephone to send a wire to General Giles, his chief of Air Staff in Washington. If the women had not taken off yet, they were not to do so. "No women," he ordered, "to fly transoceanic planes until I have time to study and approve."

Thus, women would be allowed to join their male Air Transport Command ferrying division colleagues "in every plane that's made," but not "to every fighting front." Had Nancy Love and Betty Gillies successfully delivered their B–17 to Britain, an important precedent would have been set which could have changed the course of women's utilization as ferry pilots, and perhaps affected thinking in the years to come about women in combat zones.

Oliver La Farge, who served as the on-the-scene Air Transport Command historian, suggested in *The Eagle in the Egg*, his tome on the ATC published in 1949, that General Arnold politically did not dare confront the prospect of women's flying combat airplanes so near the enemy. It was also posited that Jacqueline Cochran, as her office was in the Assistant Chief of Air Staff's, knew of the mission and somehow interfered. Indeed, when Love and Gillies returned despondent to their home bases, rumors circulated within the ferrying division that Cochran wanted to remain the only woman to ferry a bomber to Britain. As sensitive as Cochran was to the public's attitude about her women pilots, she no doubt wanted to avoid the horror of a WASP's being shot down, as many assumed had happened to the famous British ATA pilot Amy Johnson Mollison.

But the sequence of events seems to indicate that stopping the women's B–17 flight was General Arnold's decision. The AAF Commanding General acted on the assumption of the era, that women flying overseas was a preposterous idea, if not dangerous—not to their safety, but to the morale of his combat-ready troops. Underlying this attitude was a fear of scandal. A case in point was the barring of women from flying pursuits to Alaska.

Throughout the war, WASPs picked up the pursuit planes fresh off the assembly line at the Bell Aircraft factory and nurtured the new engines all the way to Great Falls, Montana. But at Great Falls, women had to hand their new airplanes over to male ferry pilots to fly the ferry route to Fairbanks, Alaska. At Fairbanks, a squadron of Russian ferry pilots transported the fighters across the Bering Strait to their war-wracked homeland.

Nancy Love repeatedly argued at Air Transport Command ferrying division headquarters in Cincinnati that women should be allowed to continue on to Alaska. To give up the planes at the United States–Canada border was not only discouraging to the WASPs, but illogical. Such a transfer needlessly occupied the time of a whole new set of pilots. And Love, of course, was right. Nevertheless, she was told, as were the WASPs themselves, that the AAF could not bear the thought of one of their women pilots being downed by engine failure or weather, stranded without hope of rescue on a blizzard-ravaged glacier.

Since the AAF had no qualms about assigning WASPs to fly red-lined aircraft on target-towing missions for a battery of untrained flak gunners, their safety was hardly the AAF's concern. In the winter of 1944, WASPs based at Romulus, Michigan, who ferried most of the Russian lend-lease fighters, found out the true sentiments of their men commanding officers. A communication from AAF headquarters in Washington, seen on the desk of their Romulus base commanding officer, definitely vetoed WASPs ever flying to Alaska, but no mention was made of fearing for their lives flying over the frozen wastelands. Rather, AAF headquarters was far more worried about the consequences of sending a woman into Fairbanks. According to the memo, the men stationed there and elsewhere in the territory of Alaska had not seen a woman of their own race for over two years.

This interpretation of projected events, as unflattering to the valiant men of the AAF as it was to the WASPs, was sufficient reason among men commanders to deny the ferrying route to the WASPs. Yet the fear of scandal still did not come to grips with the AAF's adamant rejection of women fliers ferrying "overseas" or to destinations considered "war zones." At all-male bases in the United States which did not have nurses' quarters, WASPs were billeted along with men in BOQs where temporary barriers erected between rooms were the only separations keeping the presumably ravenous young men and the equally voracious women apart.

War zones were strictly off-limits to operational women pilots. A war zone was considered *terra sacra*, an inviolable and exclusively masculine domain. Unless she was administering to the sick and dying, a woman simply did not belong there. She could not be trusted. Women pilots, though proving to be helpful support at home, where theoretically men did not want to be, were considered neither fit nor good enough for the really crucial missions of the war.

In spite of General Tunner's best efforts, WASPs were now barred from the ultimate stage of his on-the-job ferrying division training program. Consequently, they no longer fit anywhere in the systematic transition from Class I to Class V instrument-rated, overseas ferry pilot. In fact, every cockpit occupied by a woman robbed a male pilot who could go all the way through the system—to fly every plane, over every route, to every fighting front. Every woman pilot, whom Tunner had needed so badly in June 1942, was now, in September 1943, a handicap to the ferrying division, through no fault of his own. It is no wonder that he was moved, in his autobiography, *Over the Hump*, to call women pilots: "The greatest continuing hassle in the Ferrying Division, one which caused ripples of anger, frustration, and indignation . . ."

In December, 1943, General Tunner issued a quota of no more than fifteen graduates from Avenger Field a month to be assigned to the ferrying division. Trainer aircraft production had been cut back, while manufacture of more sophisticated combat planes had risen to new highs, and Tunner had to get these planes to war. In Tunner's on-the-job training system, there was nowhere for women to go; neither did it provide them anywhere to stay. There *was*, however, one exception that did not fit into the orderly progression from Class I through Class V, four-engine, overseas rated ferry pilot. These special, temperamental airplanes demanded a special type of pilot, carefully groomed by the AAF for combat, but hard to find in the Air Transport Command. As 1943 drew to a close these airplanes, rolling off the nation's assembly lines faster than any others, were needed immediately for shipment overseas. They were pursuits.

NANCY BATSON's pioneer class completed training at Palm Springs on January 1, 1944. From then on, WASP ferry pilots were assigned continuously to Air Transport Command pursuit school, first there and then to a new and expanded pursuit training center at Brownsville, Texas. Few WASPs washed out of pursuit training (one did, not for flying deficiencies, but because of severe Gulf of Mexico sunburn). But more significant, WASPs' accident rate while learning to fly America's most temperamental, high-performance airplanes was far lower than men's.

While combat pilots in the European and the Pacific theaters were thrilling at, and thanking heaven for, their fighter's fast attacks and getaways, the 117 WASPs who flew these high-performance airplanes across the country were among the first Americans to get an inkling of the new air age.

After graduation from pursuit school in Brownsville, in March 1944, WASP Margaret Kerr went to the ferrying base at Dallas and reported to the flightline eager for her first delivery orders. To her delight, the operations officer gave her a shiny new P–51 Mustang to fly to Newark, New Jersey, the ATC's port of embarkation for the European theater. Gazing admiringly out the operations office window at the Mustang, she plotted her course to Newark so that she could buzz her home town of Ada, Oklahoma. She took off and headed toward her native state. At the time she calculated to be over Ada, she began scanning the terrain below. To her amazement, she found she was already at Tulsa. She was flying so fast that she had missed Ada by 100 miles.

Other WASP pursuit pilots experienced the same bewilderment watching their country shrink below them. On a routine ferrying delivery flight in a P–51 from Long Beach to Newark, WASP Jean Landis flew from coast to coast in only nine hours and thirty-five minutes in the air, two hours faster than Jacqueline Cochran in her P–35 when she won the Bendix Race in 1938. For airline passengers, such a trip still took two, and often three, days. Another WASP, Jill McCormick, at the controls of a P–47 en route from Long Island to North Carolina, kept finding check points ten minutes ahead of where she thought she should be. When she recomputed her ground speed, she clocked 401 miles an hour, faster than the prewar international speed record.

For the WASPs, many of whom had never left their home towns before the war, to travel coast to coast two and three times a month was as exciting as around-the-world trips are for their grandchildren now. Pursuit pilot Helen Schaefer, formerly a bank worker in Cincinnati, buzzed every state capitol in the nation in her sixteen months as a ferry pilot. On her first trip into Newark in a P–51, she could not resist sightseeing. As a baffled voice called from the tower, "P–51, where are you? We no longer have you in sight," Helen flew off to buzz the Statue of Liberty.

But ferrying trips in fighters were hardly joyrides. Their charges were so precious that WASP pursuit pilots were trained to fire .45-caliber pistols. Many pursuits bound for overseas carried top-secret equipment—cameras, gunsites, and IFF (If Friend or Foe) sensors and transmitters. In P–47s flown from Republic Aviation's Farmingdale factory to Newark, medical packs stashed in the cockpit contained morphine, already a hot blackmarket item. The pistols were not meant for self-defense, however. WASPs were shown a spot on the fuselage at which they must shoot, should they be forced down under suspicious circumstances. One accurate bullet would blow up the entire airplane, rendering it and its top-secret cargo useless to a domestic spy. Some WASPs loved carrying guns and a few even achieved sharpshooter ratings. But others, like WASP Helen Schaefer, were uncomfortable about loaded arms. When she checked out her .45 at the ballistics office at Dallas, her ferrying base, she put it under her parachute seatpack and dropped the cartridges in her purse.

Once in the air, however, WASP pursuit pilots had only their skill, stamina and composure for protection. When they signed for a fighter fresh from the assembly line at the North American, Bell or Republic factory, its engine had run for an hour on the ground, but the airplane had never before been flown.

There was a special feeling while flying a brand-new fighter to an embarkation center where it would be placed in a ship bound for a combat pilot and future glory. But taking off in a new airplane, with a new engine, the potential dangers were as countless as the mechanisms under the hood. Five WASPs crashed to their deaths in pursuits, before the planes ever reached combat zones. The momentum of production was at a fever pitch in 1944, with factories working day and night throwing out up to fifteen fighters in each twenty-four hour period. On every maiden flight, whether in a new P–51 across the country or a P–47 across the Hudson River, women served as test pilots and guinea pigs for the men in Fighter Groups overseas who would be next in the cockpit. And the first entry on a pursuit's Form One was not always "A-OK."

In the spring of 1944, WASP Barbara Russell picked up a new P–51 at Long Beach, to fly it to the East Coast embarkation center at Newark. To break in the Packard-Merlin 1,695-horsepower engine, she would "slow-time" it, or keep her air speed down to 300 miles an hour for her first day en route. After she took off, she would test the instruments, navigation equipment and fueling system. If she found anything wrong in the fighter, she wanted it fixed before a combat pilot flew it on a sortie over the Low Countries.

Since her P–51 burned a voracious sixty gallons of fuel an hour, each of her three 85-gallon fuel tanks was full. First she would burn off the fuselage tank, then, switching every twenty minutes, she would alternate between the left and right tanks. As fuel weighed around six pounds per gallon, she carried over 500 pounds of gasoline in each wing. If she did not burn them down evenly, the precariously balanced plane could cause her a lot of trouble.

Barbara took off from Long Beach and climbed steadily to clear San Jacinto Mountain southwest of Palm Springs. After flying about forty-five minutes, as she crossed the Colorado River into Arizona, she switched the gas lever from the fuselage tank to the left wing tank. Immediately, the engine began to cough. Then it stopped. For a moment of dead silence, Barbara watched in horror as the air speed indicater dropped, and the Mustang hung suspended at 13,000 feet above the rugged mountains. Instinctively she reached down and switched the gas lever to the right tank and waited. The engine sputtered and caught. Her air speed increased, but for several minutes Barbara listened in a state of electric alertness to every minute variation in the drone of the engine. Her mind was racing.

If the gas connection held, she knew she could make it to Phoenix, only 150 miles away. But if it did not, between here and Phoenix there was not a landing field in sight. If anything should happen she would have to bail out. Her eyes raised to the latch that released the hatch cover. In her mind, she rehearsed what she had been taught in pursuit school. She would release the hatch. The canopy probably would be torn off of the plane by the force of the wind. She must move fast before the plane plummeted. Dragging after her the awkward parachute seatpack and holding onto the panel so as not to be blown away, she would climb out of the cóckpit. Then she would slide out onto the far edge of the wing, to avoid the propeller when she slipped off. . . . At this point in her mental rehearsal she looked down at the desolate, mountainous terrain below and decided such a plunge was not an appealing possibility. She had to make it to Phoenix.

Every minute, the P–51 burned off another pound of hundred octane from the right wing, and, to compensate for the uneven weight, Barbara pushed the left rudder pedal down with increasing force. Her leg began to throb with strain. Suddenly, she saw Phoenix in the distance. She called the tower and asked for an emergency landing, and concentrated all her attention on keeping the runway in a single spot in her windshield. As she went through her prelanding check list, her rudder pedal foot was all the way to the floor, and as she approached the runway, the plane was listing so badly, she could barely hold the nose pointed along the center line. By sheer willpower, she made a three-point landing. Only when she taxied to the hangar did she dare look at the gas gauge. The right wing tank was almost empty. "Better iron the chinks out of this airplane before some lieutenant is being attacked by a bunch of Messerschmitts," Barbara joked weakly with the ground crew as she sat and waited for life to return to her left leg so she could climb out of the cockpit. In spite of the terrifying hour she had just spent, she could not help feeling just a little bit proud.

But Barbara Russell was lucky. On April 3, 1944, an engine that sputtered and quit proved fatal to twenty-three-year-old pursuit pilot

Evelyn Sharp, as she took off in a P–38 Lightning twin-engine fighter from an air base outside of Harrisburg, Pennsylvania. As a civilian, Evelyn did not qualify for Army death benefits. The WASPs at the nearby ferrying base of New Castle, Delaware, took up a collection and sent Evelyn's family in Ord, Nebraska, over $200 to help with funeral expenses. The WASP squadron commander at New Castle, Betty Gillies, insisted that there be a uniformed representative at Evelyn's funeral and arranged for a train ticket. She asked Nancy Batson, who, like Evelyn, had come to New Castle as a WAF eighteen months before, to accompany Evelyn's body home to Ord.

As Nancy jostled along the railroad tracks past miles of Nebraska farmlands, she gazed sadly out the dusty window of the old train and let rare thoughts enter her mind about the precarious war duty she and Evelyn had chosen. In the four months she had been flying across the country in the single-seat of a pursuit plane, Nancy had never felt so lonely as she did in that crowded, antiquated train. She was suddenly overwhelmed by thoughts of death. Though only twenty-three, Nancy had developed an instinct for it. Ever since taking her first ride in a Ford Tri-motor her freshman year at the University of Alabama in 1938, flying had been everything to her. And flying AAF pursuit planes for Nancy was the most exhilarating, rewarding existence conceivable. Her life was in suspension, between taking off and landing. Every second in the air, however, Nancy was tuned into the constant, menacing hum of danger. Since arriving in New Castle to join the WAFS, Nancy had not written a single letter. She had no sense of her future beyond the war. She lived with the knowledge that if her number came up, it came up, just as it had for Evelyn Sharp, lying in a wooden casket in the baggage car.

Death had to be a matter of luck. Evelyn Sharp had been one of the best pilots Nancy had ever seen. She had even died proving it. A minute after taking off in the twin-engine P–38, at 700 feet, Evelyn had lost one of the 1,325-horsepower engines. Almost anyone would have spun in seconds later, as the torque of the remaining engine flipped the plane over to plummet dizzily to the ground. But Evelyn had jammed the rudder pedal down and had been able to keep the nine-ton fighter level. Then she had managed to bank the Lightning 180 degrees back toward the airport, and in an almost impossible maneuver, had pancaked the P–38 for a wheels-up landing downwind. The plane was hardly damaged. But the nose of the P–38, slightly ahead of the two engine nacelles, had hit the ground first. The steering column rammed straight through the cockpit, and Evelyn had been thrown through the canopy, breaking her neck. If the engine had quit with five more feet of altitude, Evelyn would have saved her own life.

Suddenly, Nancy was jarred away from her thoughts as the old train screeched to a halt at Ord. Seeing her uniform, a man on the platform approached her and introduced himself as the town undertaker. As they waited for the coffin to be carried off the train, he explained that Evelyn's family wanted an open casket, and had asked him to prepare her body before they saw her.

As Nancy sat in the waiting room of the funeral home, waves of horror rose in her as she thought of Evelyn's beautiful face, so alive and laughing, now shattered beyond recognition. For an open casket, the undertaker would have considerable cosmetic rehabilitation to do.

Late in the afternoon, the coffin was wheeled out into the waiting room. Soon the Sharp family arrived and for two hours, Nancy stood by while they vented their grief. Then other townspeople began pouring into the funeral home, until it was packed with people weeping. Nancy was amazed at the effect Evelyn's death was having on the town of Ord. A man, spotting her uniform, asked Nancy if he could drape Evelyn's coffin with an American flag for the burial. The remark stung her. Neither Evelyn, nor she—nor any of the WASPs—were really military. They flew with no insurance, no benefits—and no honors if they died. Nancy suddenly was furious at the injustice that this fine young pilot who died for her country should, as far as the Army was concerned, be buried as if she had anonymously been hit by a truck. Nancy took the man's hand and said, "Of course," thinking they should bury Evelyn Sharp like the military heroine she was.

The next morning, the town of Ord honored WASP Evelyn Sharp at a two-and-one-half-hour funeral. As Nancy listened to the eulogies, she realized that her friend always had been a heroine to Ord. An orphan, Evelyn had been adopted by the Sharp family. The vivacious teenager had taught many of the children of the town how to swim and ride horseback. At fourteen, Evelyn had discovered airplanes. By the time she was eighteen, she had earned her air transport license, and the businessmen of Ord raised enough money to buy their favorite daughter an airplane. She had flown in the town's first sack of airmail letters. Before joining the WAFS, Evelyn had thrilled the town with her barnstorming stunts and taught many of its sons, now serving in combat, how to fly. As the service drew to a close, the mayor of Ord told the congregation that hereafter, the town airport would be known as Evelyn Sharp Field.

After the funeral, the throngs moved slowly to a cemetery on the outskirts of town. Many townspeople approached Nancy tearfully, to tell her how much they had loved her friend, and how proud they were that Evelyn and she were serving their country as pilots. No Congressional Medal of Honor winner, Nancy thought, could be honored as lavishly as WASP Evelyn Sharp. As a bugle played taps, her flag-draped casket was gently lowered into the ground. With this, Evelyn Sharp's last landing, Nancy Batson, her official duties over, finally broke down and wept.

As THE SPRING of 1944 progressed, the number of WASP ATC pursuit pilots grew to almost one hundred. They viewed their unique wartime flying missions in the nation's fastest, most temperamental airplanes with continuing awe. That so few WASPs were killed in pursuits was a tribute not only to their flying skill, but also to the care and efficiency of assembly line workers, many of whom were women. But women in a defense plant were taken far more for granted. A woman in the cockpit of the war's famous fighters caused consternation.

Fresh from pursuit training in Palm Springs, WASP Carole Fillmore was given her first P–51 Mustang to deliver to Newark. As it was winter, she flew a southern route across the country, intending to turn north up the East Coast, to avoid inclement northern winter weather. Her first day, she flew all the way from Long Beach to Athens, Georgia, where she ran out of daylight. But when she tuned in the Athens tower and called for landing instructions, no one answered. She called the tower again and still received no response. In minutes she was directly over the field. She began to circle, and asked for a radio check. Suddenly an exasperated voice rasped into her earphones: "Will the woman who is calling please stay off the air, we're trying to bring in a P–51."

Carole looked around her in the growing dusk for the other Mustang. Seeing none, she called in again for landing clearance.

"Will the lady who's trying to get in, please stay off the air!" shouted the tower. "We are trying to make contact with a P–51!"

Carole was beginning to feel weary from her long day in the air, and her patience was running out. Finally, she pressed the button on the stick activating her throat mike. "For your information, the lady who is on the air is *in* the P–51," she said, and without waiting for an answer, turned on final approach and headed straight down the center of the runway at 120 miles an hour.

The radio suddenly came alive. "Yeah, man, I hear you, I hear you, did you see the light, you're fine, you're coming in fine, just great. . . ." Carole made a perfect landing. "Aw, that was beautiful," drawled the tower.

As Carole taxied up to the flightline, scores of young cadets were pouring out of the ready room to gawk at the huge new fighter they all dreamed of flying someday. Carole parked, cut the engine and threw back the canopy. Her dark blonde hair flew up around her face in the warm breeze. The entire crowd seemed to lurch backwards in astonishment. "It's a girl!" someone shouted. As she climbed out of the cockpit onto the wing, her audience burst into cheers, whistles and waving hats. Carole stood on the wing of her pursuit plane and grinned. She felt like Lindbergh landing the *Spirit of St. Louis* in Paris.

WASP ferry pilots in their wide travels around the country got used to reactions from the sublime to the ridiculous. One ferry pilot, refreshing herself in an air terminal lavatory after a long leg of a trip, encountered a janitress cleaning the washbowls. Seeing the young woman, begrimed and wearing a bulky khaki flying suit and clumsy leather flying shoes, the janitress shook her head in amazement. "A lady plumber!" she sighed. "Girls are doing everything these days."

But a woman pilot was often viewed with suspicion. Flying a P–51 Mustang from her home base of Long Beach to Dallas, WASP Barbara Russell landed at an isolated airfield in western Texas. Leaving her fighter by the hangar, she walked into the operations building to arrange to remain overnight. When she returned to her P–51 to get her briefcase, a young GI, stern-faced and with gun cocked, was standing by the plane, on guard against miles of empty Texas plains. Barbara, its unlikely pilot, became his first intruder. It took her strongest words to persuade him that she had just flown the P–51 into the field and wanted to get her belongings from the cockpit.

Although hotels where women ferry pilots stayed and air bases where they routinely refueled received letters from the ATC ferrying division and WASP headquarters to expect women, society at large was ignorant of the WASPs, even when over 1,000 were flying for the AAF, a third of whom were with the ATC and out ferrying airplanes daily. Lacking an official uniform until the late spring of 1944, the WASPs' problems of recognition were exacerbated. In spite of their wings, the nondescript tan gabardine flying slacks and white shirt did not look very military. Even the W.A.S.P. insignia on their caps did not mean anything. One WASP inverted the letters to read "S.P.A.M." throughout a cross-country ferrying trip, and no one even noticed. One weary WASP foursome, after flying P–51s across the country from Long Beach to Newark, took a bus into New York to spend the evening. Strolling down Broadway, they found Jack Dempsey's restaurant and fought the busy evening crowd of soldiers, officers and their dates to the bar. As the four women pilots ordered drinks, they became aware that the bartender was glaring at them. "No women in slacks," he said haughtily. "But these are our uniforms!" the WASPs protested. The maitre d' was already on his way over to them, however, and the WASPs were ushered briskly out the front door.

Halfway down the block, the women heard shouts. "Ladies! Ladies!" The manager of Dempsey's was running after them. "Please come back. We profusely apologize. We didn't realize who you were. We thought you were . . . well, just women in slacks." He was huffing and puffing so badly, and the WASPs were so hungry, that they agreed to return. An officer at the bar, they learned, had witnessed the confrontation, grabbed the manager, and said, "Do you know who you just threw out? They were WASPs, women Army pilots!" The WASPs spent their evening at Dempsey's on the house.

But women pilots did not always have a champion, nor were such chance encounters with the civilian world always so amicably resolved. In the early spring of 1944, bad weather grounded four WASPs in Americus, Georgia. They left their airplanes in the hangar at the airport and caught a bus into town to look for a hotel room. No sooner had they started down the main street, when a police car pulled up beside them. Two policemen got out and demanded that they come with them to the police station. Women were not allowed on the street at night in slacks, one cop said sternly. No amount of insisting that they were Army pilots in uniform would do.

At the police station, the women were locked in a cell. They tried to persuade the sheriff to let them make a telephone call to their ferrying base. Instead, he decided to test their story by calling the commanding officer of the air base where they had landed. The C.O. was at a party. The WASPs listened to the sheriff's end of the conversation with growing alarm. "We have a few girls down here *impersonating officers*," he said. The mockery with which he enunciated his last two words showed clearly the genre of women he thought them to be. "That's exactly where I've got 'em," the sheriff said smugly, and hung up the phone.

Not until after 2:00 in the morning were the WASPs allowed to make one phone call. They decided to call Cincinnati and wake up Nancy Love. When they told her where they were, Love demanded to speak to the sheriff. The WASPs had been fighting back tears. But as they watched the sheriff's face, they wanted to laugh. He winced, and held the telephone receiver away from his ear. Belying her satiny looks, their ferrying division chief commanded a rough pilot's vocabulary. Love's threats, ranging from a lack of patriotism to a personal summons before the Commanding General of the Army Air Forces, were evidently persuasive. The WASPs were released and driven back to the airfield, where they waited with agitation until dawn to take off and fly as far away from Americus as they could.

The following day, in a state of shock that WASPs might actually be jailed, Nancy Love alerted her WASP squadron commanders that emergency twenty-four hour telephone numbers be given to all of her WASP ferry pilots.

WASP FERRY PILOTS were in such demand across the country that they often found themselves away from their home bases for several days. Aside from a pile of tech manuals to refresh their memories on all the different types and series of pursuits or other airplanes they might be ordered to fly, they traveled light to fit into the cramped cockpit. WASP Alma "Pat" Velut, who dreamed of traveling far from her home town of Granada, Colorado, had a typical system. She put her orders, maps and pilot logs in a briefcase, some make-up in her purse and a toothbrush in her shirt pocket. When she landed, she washed out her underwear and hung it on her hotel bedstead. After scrubbing her shirt collar, she put the Gideon Bible on top of it so it would dry as if just ironed. Then she laid out her slacks between the mattress and box springs and bedded down for the night. The next morning she reported back to the airport, fresh and official for her dawn departure.

Routine orders were to deliver the airplane and then return to home base by the fastest method of transportation possible, to receive their next delivery assignment. As women pilots were forbidden to hitch rides on military planes, many a night was spent asleep on their parachutes in bus and train stations across America. The most desirable method of returning to base was to be ordered to fly an airplane back. But sometimes the plane was going in the wrong direction.

Based at New Castle, Teresa James received orders to pick up a P-47 Thunderbolt from Farmingdale, Long Island, and deliver it to Republic's modification center in Evansville, Indiana. Looking forward to a quick 800-mile trip in the fast pursuit, and assured of another P–47 to fly home again, Teresa put on a pair of green WAFS uniform slacks and a regulation shirt, stuck a lipstick tube in her pocket, and took off into the summery blue sky. Around noon, she arrived in Evansville, exhilarated from her beautiful morning trip over the Appalachian mountains. Just as she was grabbing a sandwich at the canteen, she was rushed by a frantic operations officer with orders for a P–47 which had to be taken the next day to Long Beach, California. Teresa did not dare to tell him she could not go because she only had one shirt. Instead, she went to the PX and bought a toothbrush, toothpaste and a tube of cold cream.

The next day, Teresa flew almost 1,800 miles from Indiana to Long Beach. No sooner had she signed over the P–47 to operations when the officer behind the desk asked, "Have you ever flown a P–51?" Teresa shook her shock of black hair. "That one over there has to go to Fort Myers, Florida," he said, pointing toward the flightline. "Here are the tech orders. Sit in the seat and get familiar and just check yourself out in it." Teresa was delighted finally to fly the beautiful Mustang. In Wilmington, she had gleefully sneaked transition training in the P–47 that spring, before WASPs were sent to pursuit school. Teresa had not seen very many P–51s. That evening she washed out her shirt, borrowed an iron from a nurse in the nurses' quarters, and studied the tech orders of her first Mustang, which, without the least trepidation, she would hop into the next day and fly across the country.

The next morning, after twenty minutes practicing takeoffs and landings so that she felt comfortable in the new fighter, she headed east. Halfway across Texas, however, she ran into the outer fringe of a summer storm in the Gulf of Mexico. Rain covered half of Texas and all of Louisiana. For five days, Teresa could only make short hops from base to base during breaks in the overcast. Her loafers were permanently caked with mud from the soggy runways. The humidity would not allow her socks and underwear to dry overnight before she had to put them back on in the morning.

As she did not have her uniform jacket, she could not enter any of the air base officers' clubs. (A WAFS uniform, of course, could not be found at PXs.) She began to tire of sandwiches at operations canteens, and she desperately hoped that at Fort Myers, she would be given a plane to fly home to New Castle. When she signed over the P–51 to Fort Myers operations, however, the officer indicated an AT–6 out on the flightline that was due in Oklahoma. Teresa gasped.

Two days later, Teresa was in Tulsa. On arriving at the air base, she was delighted to see Barbara Erickson, who had just delivered a plane on the East Coast and was on her way back to her home base at Long Beach. Though they had been stationed together at New Castle in the early days of the WAFS, they now saw one another only when their flight paths crossed in the middle of the country. The reunion was a happy one, especially for Teresa, who had just received orders to take a P–39 Airacobra to Great Falls, Montana. She was salivating for some real meals in an officers' club, and since Barbara was going home, she offered Teresa her WAFS jacket.

The next day, Teresa flew her P–39 1,200 miles from Oklahoma to Great Falls, Montana. As soon as she walked into the operations office, she was handed another set of orders. She could not bear to open them until she had found a cup of coffee. Then she sat down and began to read. For a minute she sat in disbelief. At last she was being ordered to fly a P–47 back to New Castle.

Twenty-four hours later, Teresa stood in the door of her room in the WASP baracks at New Castle. Greeting her was a stack of mail three inches high and several WASP colleagues pointing at her and laughing. "Jamsie, you look like the wreck of the *Hesperus!*" Teresa's uniform slacks were so stretched out at the knees that they looked like bloomers. Her day's hop in a P–47 to Evansville had turned into a 4-week, 17-state, 6-airplane, 11,000-mile trip.

In June 1944, Teresa James returned from a long ferrying trip to grim looks from her WASP friends. They handed her a telegram. It was the one Teresa and millions of wives across America dreaded every day of the war. Teresa's husband, George, a B–17 pilot, was reported missing in action. The couple had said good-bye all too quickly in the fall of 1943, on a brief visit from George to New Castle before he was shipped out. He had a 48-hour leave to explain to his weeping wife why he had suddenly left his assignment as an instructor in Oklahoma and volunteered for combat. As a WASP, Teresa had officers' club privileges. But when the couple appeared at the door, she was told sternly that she could not bring her husband in. George was an enlisted man. The desperate couple had nowhere to go. The WASP squadron commander, Betty Gillies, defied possible court-martial and signed her pilot out to stay off-base. Then George was gone. A month after receiving the telegram, Teresa saw a newsphoto of an unidentified B–17 crew, which, having parachuted from their flaming Fortress over France, was being led away as prisoners of war. One of the prisoners was her husband. France was liberated in late August. Teresa flew on, in hope of some word that never came.

WHILE A fighter pilot overseas got to know every inch, cranny and peculiarity of his own airplane, WASP fighter pilots had to keep current in all of the different types and their serial variations. With each new series came innovations, and some with dangerous quirks not mentioned in the tech manuals, but rumored among ferry pilots. WASPs learned to listen hard—they might get that airplane the next day.

The day President Roosevelt was re-elected for his fourth term in November 1944, WASP Nancy Batson, taking off from Pittsburgh Airport, learned first-hand about a series of P–38 Lightnings that were notorious for a faulty nosegear. The P–38 was a formidably powerful twin-engine fighter, with two 1,325-horsepower liquid-cooled engines located in nacelles that extended back in parallel tail booms. The cockpit floated on the wing between them like a passenger cab attached between two motorcycles. As if Lockheed, the manufacturer, knew of the series' reputation, designers installed shiny aluminum reflectors on the insides of each engine nacelle so the pilot could monitor the state of the nosewheel. In the cockpit was a manual pump to serve as a backup to the hydraulic system operating the landing gear. Under the pilot's seat was a button connected to a CO_2 cartridge which could be exploded as a last resort if the pump did not force the nosewheel out of its casing under the cockpit.

Moments after Nancy Batson took off from Pittsburgh on the last leg of her trip from California to Newark, she noticed that the P–38's engine coolant needle was oscillating. Fearing engine failure, she immediately radioed the tower that she was returning so a mechanic could check out the trouble. Though many months had passed since Evelyn Sharp's fatal crash in a P–38, Nancy was still haunted by apprehension whenever she took off in one. As the tower cleared her for a straight-in approach, Nancy put the gear handle down and heard the familiar hum of the wheels. But the red gear-in-progress light did not go out. Glancing in the aluminum reflectors, she saw that the nosewheel was dangling, only halfway erect, under her. She immediately began to work the manual pump.

After half an hour, the pump pressure had still failed to budge the nosewheel. At the instruction of the tower, she flew out over the city away from the airport and climbed to 8,000 feet. Then she dove the

P–38 as sharply as she could and pulled up suddenly, to try to force the nosewheel out by centrifugal force. The roar brought the good citizens of Pittsburgh bursting out of their homes to look up into the sky. And in the air, every plane for miles heard the running exchange between the tower and the woman pilot with the Southern drawl: "How are you doing?" "I'm still flyin' over Pittsburgh and still pumpin' and still have a red light." At the airport, a crowd of Red Cross workers, newspaper reporters and photographers had gathered. An Air Transport Command officer in the vicinity was now on the tower radio with tower operators. All were anxiously waiting for the P–38 to get its gear down and land safely. After a while, the officer called Nancy Love in Cincinnati to tell her that one of her ferry pilots was in trouble. She asked who it was. When the officer told her she was Nancy Batson, Love said not to worry, Batson would do fine.

Love's assurances did not help Nancy. For two hours, she had pumped the manual pump and dived her P–38 down at the city of Pittsburgh. Her engine coolant needle had regularized, but she was now getting low on gas. Until then she had been too busy to worry. But with the possibility of running out of gas before the nosewheel dislodged she experienced her first moment of fear. A brief vision of Evelyn Sharp being thrown from the cockpit as she landed her P–38 wheelless flashed through Nancy's mind. "My time hasn't come," she said to herself, and was somehow sure this was true. She decided to shoot off her last resort, the CO_2 cartridge. Mustering all her remaining energy, she dived the P–38, pulled up sharply, pumped the manual pump and set off the cartridge. It exploded with such force that Nancy thought she must have blown the nosewheel right off. But when the smoke cleared, the red light had gone out. In the shiny reflectors she saw the nosewheel pointing straight down.

There was great rejoicing at the Pittsburgh airport as the P–38 touched down on the runway. When Nancy caught her breath, she was aware of jeeps, ambulances and firetrucks converging on her from every taxiway. As she rolled to a halt, a jeep caught up with her and an officer jumped out and up onto the wing next to the cockpit. He looked wild-eyed and was shaking with concern. "G-g-get out!" he stuttered, "I'll take it in." Nancy, who had no idea of the frenzy she had caused on the ground, looked at him in astonishment.

"*I'll* taxi it in," she said. "I got it this far—I'm not fixin' to get out at this point."

* * *

DURING THE fair-weather summer months of 1944, WASP Air Transport Command ferry pilots delivered more airplanes, many of them pursuits, than in any preceding months—1,280 in July and 1,049 in August. The WASP squadron at Long Beach, the largest ferrying base, was now eighty strong, though the women ferry pilots were rarely at their home bases for a head count. But since the spring, a small squadron was busy making one of the WASP's most startling contributions to the domestic war effort. Eight WASP pursuit pilots, stationed at the Republic Aircraft factory in Farmingdale, Long Island, and at Republic's Evansville, Indiana, factory and modification center, were delivering every single P–47 Thunderbolt fighter that came off the assembly line. Every Thunderbolt shipped to Britain, and later in the fall, to liberated France and Italy, had a WASP in the cockpit during the first leg of its journey.

On Tuesday, September 20, 1944, the WASP P–47 squadron was formally honored at a ceremony, at the Republic factory on Long Island, celebrating the 10,000th Thunderbolt produced in only two and a half years. Jacqueline Cochran, who had flown its prototype, the P–35, to victory in the 1938 Bendix Race, smashed a bottle of champagne against the P–47's cowling, christening it *Ten Grand*. Then, as newsreel cameras whirred and thousands of Republic factory workers cheered, into the cockpit hopped WASP Teresa James, who by drawing the longest straw in the squadron, had won the honor of flying *Ten Grand* on its first leg to glory.

BY THE fall of 1944, half of all the ferrying division's fighter pilots were women, and three fourths of all domestic deliveries of America's fastest, most exciting airplanes were accomplished with WASPs in the cockpit. In their unique place in the ferrying division, the WASPs' contribution was finally felt in the military organization they were first called forth to serve.

XIV

So Easy Even a Woman Can Fly It

ON MARCH 2, 1944, Jacqueline Cochran presented to General Arnold a complete summation of her WASP program to date, upon which he could base his testimony before the House Military Affairs Committee on a bill to bring the WASPs formally into the Army Air Forces. Cochran was now exuberant. She had the facts and statistics. Her experiment was a success.

It had been a year and a half since the original twenty-eight professional women pilots appeared one by one at the guardhouse at New Castle Army Air Base, Delaware, to join the Women's Auxiliary Ferrying Squadron. Sixteen months had passed since the first twenty-eight Woofteds reported to the motley flightline at Houston Municipal Airport to start the Women's Flying Training Detachment. Now in March 1944, 541 women pilots had graduated from Jacqueline Cochran's training program, and almost 500 were currently in training at the all-WASP air base, Avenger Field, Sweetwater, Texas. And sitting on Cochran's desk were approved applications from over 1,200 qualified women pilots waiting to fill future WASP classes. Her school could easily be expanded by fifty trainees a month, Cochran wrote Arnold, and if he approved, she could provide the AAF with 2,554 WASPs by January 1946.

When Cochran was named Director of Women Pilots in the Army Air Forces nine months before, there had been sixty-eight women pilots, most of them ferrying small trainers for the Air Transport Command. Now WASPs were assigned to almost every air force and command that had domestic flying operations, performing their missions to everyone's satisfaction.

A special group had graduated from four-engine bomber school in January 1944, and were now assigned to train B–17 turret gunners at a Florida air base. Other WASP bomber pilots were with various AAF tow-target squadrons in the First, Second and Third Air Forces.

One hundred eighty WASPs were now stationed with the AAF Training Command as test pilots at repair depots, tow-target pilots for pursuit gunnery practice, and training-staff pilots at navigator and bombardier schools. In the future, Cochran suggested that when the WASP became militarized and protected by military insurance and hospitalization, as part of the AAF, women pilots could serve in all theaters of war, except for ferrying that would take them directly into combat zones.

For General Arnold, Cochran's report and recommendations talked sense. In early March, his Army Air Forces were making critical moves. In the battle raging since January against German divisions on the beachhead at Anzio, Italy, the Germans finally appeared to be on the defensive, and the invasionary forces were beginning to pound inch by inch toward Rome. But the advance was excruciatingly slow. German installations were so entrenched in the mountains south of the Eternal City that it would take three months to go those crucial and torturous thirty miles. On March 4, the first American bombers with their fighter escorts headed for Berlin, to join the British Lancasters and Hallifaxes in a relentless rain of destruction on the capital of Nazism. And the build-up for a massive invasion of Europe, to be commanded by General Dwight David Eisenhower, was well underway. Hundreds of thousands of American troops were pouring into Britain. Meanwhile, air raids were being planned for May on German airfields in northern France. Every week hundreds of AAF combat pilots were getting their orders to ship out for overseas, and as far as General Arnold was concerned, every woman pilot Cochran could give him would help keep his domestic war machine performing at top speed.

In January, the War Department had announced that women pilots' fatal accident rate was lower than that for men flying domestically for the AAF. Their nonfatal accident rate was lower as well. But General Arnold was also personally indebted to the WASPs; because of them, his own son, Lieutenant W. Bruce Arnold, a twenty-four-year-old artillery officer, was able to go overseas.

After a less than enthusiastic reception at Camp Davis, North Carolina the summer before, the task of training gunners had, in six months, become one of the WASPs' fortes. In fact, by the spring of 1944, they were giving target practice to every type of gunner in the AAF.

As the Allies mobilized for the invasion of Europe, vast seas of tank units were training for the ground operations which would push through France, Holland and Belgium toward Germany. One important phase of training involved rough-terrain maneuvers for tank gunners. At Camp Irwin, the armored vehicles were half-tracks, propelled by continuous chain tracks like tanks but supported by two front wheels. Fifty-caliber guns were mounted on the back of each half-track. In order to pass the rough-terrain requirement, gunners had to hit an aerial target a specified number of times while the half-track was careening over wide-open terrain at forty miles an hour. For the fledgling gunners, the maneuver was like firing from the back of a wild steer. Their guns had no settling mechanisms, and as the half-track dipped down a gully or climbed a rise, the gun bores dipped and rose with the vehicle. Until the gunners got into the swing of things, shots tended to be rather wild. After one day towing targets, male pilots stationed at Camp Irwin refused to fly any more. They were not required to fly in any situation domestically that endangered their lives, they claimed. In desperation, Lieutenant Arnold and the other artillery officers at Camp Irwin tried to have the rough-terrain maneuvers waved, but high command refused. Mobile unit gunners, drivers and their officers, anxious to get overseas, were stymied.

Then one morning, the mobile units were ordered out into the desert. From the direction of Camp Irwin came an airplane towing a target. The half-tracks began to roll across the rough ground and the gunners began joyfully to shoot at the target. After half an hour, Lieutenant Arnold heard his ground-to-air radio crackle and then, to his amazement, through the earphones, came a female voice. "Three holes in the tail, boys, that's a little too close!" The plane swooped down over the half-tracks, dropped the muslin target, and zoomed back toward Camp Irwin. Shortly thereafter, the plane was back with a fresh target. The young artillery officers were all incredulous. Lieutenant Arnold's illustrious father certainly had some unorthodox answers to winning a war.

Every month, Jacqueline Cochran was pinning wings on fifty new WASPs, who fanned out from Avenger Field to air bases across America. Her mid-March graduating class was the tenth. Even so, the domestic flying establishment was so vast, and new assignments for WASPs so far afield, that women pilots were still being met with astonishment and incomprehension as if the program were new. When WASPs Margaret McNamara and Marilyn Seafield reported for duty in January 1944 at Craig Field, an advanced, or AT–6, training base in Selma, Alabama, the commanding officer was so startled he fell out of his chair. At another training base, where fighter pilots in the last phase of their combat training learned to shoot the .50-caliber machine guns in the wings of their P–47s, Dolores Meurer and two other WASPs were assigned as tow-target pilots in February. Their first week on duty, Dolores sat down at a table in the officers' mess with several fighter pilots and the base commanding officer. No sooner had she begun eating when a waiter appeared behind her chair and told her the colonel requested she sit at another table. Flushed with embarrassment, Dolores ran out of the room. "I will not have a woman at my dinner table," the colonel told men pilots around him. "It's unmilitary."

Indeed, the pattern set over the past year and a half continued. While in the nurturing atmosphere of a training situation, the WASPs were accepted and amply able to prove their flying ability in the largest bombers and fastest fighters. But once they set forth from training bases to duty assignments on operational AAF bases, they were repeatedly met with distrust. The most flagrant instance occurred when eight of the "Lucky Thirteen" celebrated WASP graduates from B–17 school at Lockbourne Army Air Base reported for duty at Buckingham Army Air Base, Fort Myers, Florida. (The others had been asked to remain at Lockbourne as engineering pilots.)

Buckingham was a gunnery training base, which concentrated on teaching B–17 turret gunners to defend the Flying Fortress. Buckingham's male B–17 pilots who flew training missions were usually in transition, either to duty overseas, or to further four-engine training in B–24s or, later, in B–29s. A permanent staff of women Fortress pilots was a logical addition to the transient male squadron. When the ladies of Lockbourne arrived, however, including four of Lieutenant Mitchell's meticulously trained students, Buckingham's Director of Flying would not let them near a Flying Fortress. "I cannot take responsibility for a woman flying a B–17," he stated flatly. Instead, he sent them off to fly AT–6s on dive mission for ground gunners' tracking practice.

The WASPs were heartbroken and incredulous as they gazed longingly at the "Big Friend" basking on the flightline in the warm Florida sunshine. After a month flying AT–6s, they mustered their courage and telephoned Jacqueline Cochran in Washington. A few days later, the Director of Flying relented. The WASPs would be allowed to fly as B–17 copilots, but only if they passed check rides in the Fortress. Remembering the many grueling hours with Lieutenant Mitchell, Frances Greene and Charlotte Mitchell thought the check rides would be a snap. But the Director of Flying made the WASPs check out for an entire week. They were given not one but several check rides, lasting two to three hours each, and not for a minute were all four engines going at once.

The better the WASPs did on two- and three-engine procedures, the harder the Director of Flying pushed them. They felt as if they were being tortured by the enemy. "I just can't figure out how you girls can be so strong!" he shouted in exasperation. Finally he had to pass them all. He did not know that back at the WASP wing in the nurses' quarters Lieutenant Mitchell's star students had circulated the trick he had taught them of hooking the free foot behind the unused rudder pedal to give leverage and spell the throbbing rudder pedal leg doing all the work.

Although some of their receptions by base commands continued to range from startled to downright chilly, WASPs were being warmly embraced by understaffed air base maintenance officers across the country, whose telephones rang constantly with urgent requests for pilots to flight-test airplanes out of the repair shop, pick up engine parts and transport air base personnel. To them, WASPs were a godsend.

Since Pearl Harbor, Barton K. Yount, Commanding General of the AAF Training Command, had built up a prodigious and massive training effort on air bases across the country. The 318th Flying Training Detachment at Avenger Field was only one of several hundred on which AAF cadets, fighter pilots, bomber pilots and flight crews of gunners, navigators and bombardiers were trained. It took eighteen months' preparation on the home front before a Flying Fortress crew could drop a single bomb or fire a shot at the enemy overseas. With every available combat pilot being shipped out for Europe or the Pacific, General Yount was hard-pressed to maintain a competent staff of pilots assigned to the Training Command. Yet these pilots were the backbone around which Yount had to form a fighting air force.

For General Yount, the matter was even more muddied by the fact that combat-trained male pilots sorely begrudged domestic flying jobs. Though the Training Command wanted its recent cadet graduates to have some domestic flying experience to build their confidence in the air before they were sent to meet the enemy, the men, ready for combat duty, did not want to stick around flying a bunch of fledgling navigators over Texas, or flight testing trainer aircraft that had been in the repair shop. Such flying jobs were risky, with no payoff in glory. The AAF did not blame its top-flight men pilots for shunning such duties, nor could it afford to keep combat-trained pilots flying domestically for long. Yount was delighted with his growing corps of WASP staff pilots, and with the efficiency and enthusiasm women were bringing to his training effort.

By the spring, women pilots were flying planes from which bombardier trainees dropped live bombs on targets in the desert. Others stationed at navigator training bases were taking directions from inexperienced future bomber crew navigators who were training in overwater hops far out into the Gulf of Mexico. More than once, a WASP pilot had to swallow hard and alert navigators in the belly that if they did not turn back, they would all end up in the drink, out of gas. Relations between women pilots and their charges were not always amicable. Navigators and bombardiers were often washouts from AAF cadet school who viewed successful Sweetwater graduates with less than good humor.

But there was one job in the Training Command that men pilots often simply refused to do, one that was crucial on every training base across the country—that of test pilot. On Yount's basic and advanced cadet training bases there were hundreds of airplanes that had to be maintained at a high standard of safety because they were flown by the war's future combat pilots. Men AAF pilots were far from eager to take on test pilot duties, which they perceived as a crazy risk for such an ignominious wartime task. If they were going to risk their lives, they wanted to do so in combat. And Yount agreed with them. During 1944, WASPs on over fifty AAF training bases certified to the safety of trainer airplanes flown by thousands of male cadets and their instructors. On many training bases, WASPs took over all of the test flying missions.

For WASP test pilots, every flight was a deadly serious matter. Every airplane they climbed into had had something wrong with it, from a broken propeller to an exploded engine. Every second, from the time they started the engine until they taxied back to the hangar, they had to be alert to any possible deviation from normal in the planes balance, aileron and rudder movement, or the sound of the engine. Being a test pilot was a lonely responsibility.

Maintenance units were overworked at busy training fields, and they depended on the keen observation of test pilots to safeguard against oversight. When WASP Gene Shaffer reported to Gardner Field, an AAF basic flight training school near Bakersfield, California, in April 1944, she learned that the maintenance officer did not have time to explain to her what repairs had been made. The Form One of her first BT–13 out of the repair shop read "Replace right wing." As there obviously were two wings on the BT, Gene knew the job had been partly done, but as to what had been wrong with the wing, she had no clue. A clipboard was thrust into her hand, with a testing procedure check list. Gene climbed into the cockpit, put the clipboard on her lap and took off. The first test in a newly fixed airplane was to reduce the power and maintain the slowest possible airspeed that would keep the BT aloft. In "slow-flight" the least imbalance or snag in the aileron or rudder movement would be evident to an experienced pilot.

Gene Shaffer knew the BT–13 better than most Avenger Field graduates. In February 1943, after their graduation with the class of 44–1, Jacqueline Cochran had sent Gene and fourteen other WASPs to Randolph Field, in San Antonio, Texas, known among pilots as the "West Point of the Air," to attend AAF basic flight instructor school. Gene and ten of her classmates graduated with flying colors. But resistance had been strong to the idea of women flight instructors at cadet training bases, and none of the WASPs was ever given a male cadet to instruct. They remained merely an experiment to find out if women could pass the course. But Gene was soon grateful for the advanced acrobatics training she had received in the BT–13.

After several minutes of slow-flight testing the BT–13's new wing, Gene was now to do some stalls. As she pulled up the nose and began to feel the first shuddering indication that the wings were about to stall, the right wing rapidly sank and the plane began to tip sharply to the right. Gene gunned the engine and pushed the nose down quickly. She marked on her check sheet, "Loses 1000 feet toward right wing on stall."

The next procedure was to climb up to 10,000 feet and throw the bulky trainer into an intentional three-turn spin. Gene did not like the idea of spinning a plane that had just had a wing replaced. But she was the test pilot, she reminded herself. She had to see if the plane would stay together in the novice hands of a cadet. She climbed high above Bakersfield, pulled up the nose, and fell spiralling down toward the valley floor. As she spun, Gene looked apprehensively out at the new wing. Rivets were popping up and down in their holes with the stress of centrifugal force—hardly a comforting sight. After her third turn, she popped the stick forward and pulled the BT level again. "If the wing stayed on after all this I suppose any cadet can fly the crate," Gene assured herself and banked back toward the base. On the ground, she met the dozen other WASP test pilots of Gardner Field. "You wouldn't catch an AAF officer doing this job." laughed Catherine Murphy, a six-foot WASP from Minnesota, who had learned to fly in a Cub on skis, as they all walked back to their barracks at the end of the day.

Yount was so pleased with the eagerness and competence of the WASPs to fly any of the missions so maligned by men pilots that he began to send them on his most hazardous test pilot assignments.

In the summer of 1944, a basic training base in Greenwood, Mississippi, was hit by a hurricane which damaged almost every airplane on the field. As maintenance unit mechanics began working around the clock to repair the planes so that training on the field could resume, Yount asked Jacqueline Cochran for some WASPs to be sent to Greenwood as a crisis squad to flight test the scores of repaired BT–13s. One of the WASPs was Gwen Clinkscales, a biologist from Rocky Hill, North Carolina.

For two weeks, Gwen and the other WASPs tested up to ten BTs a day in the stiflingly humid summer heat. Late one afternoon, Gwen was given a basic trainer which had hit another plane in the high hurricane winds, tearing its right wing off the fuselage. Gwen walked around the plane, meticulously checking rivets and seams. The wing looked tight as a drum. She taxied out to the runway and gave the BT full power for takeoff. Just as it broke ground, Gwen felt the plane pull violently to the right. She immediately jammed down the left rudder pedal and threw the stick to the left, but the BT still felt as if it were about to cartwheel across the field. "This could be it!" she thought as her heart pounded furiously. She had no idea what was wrong with the plane. Forcing the stick to the left and hanging on it with all her weight, Gwen realized that the BT was banking itself to the right. She let it turn and when she was directly over the runway, she cut the power, pushed the nose down and fell with an awkward thud onto the concrete. Wiping the sweat from her forehead, she taxied slowly back to the hangar and told mechanics what had happened. They scratched their heads and wearily pushed the BT back into the repair shop. The next day when she reported to the hangar, she asked what had been wrong with the imbalanced BT–13. A mechanic approached her sheepishly. Without realizing it, he said, one of the guys had riveted his fifty-pound tool chest into the wing.

To be contributing to the AAF was its own reward. But the most important event since graduation to bolster the WASPs morale and self-confidence was about to occur—the arrival of the official WASP uniform.

When it came to uniforms for her WASPs, the Director of Women Pilots in the AAF was not to be outdone. Fashion and beauty had been her business for over twenty years, and her WASPs were going to look *smart*. Late in 1943 the AAF offered her two options: to adapt uniforms fabricated and subsequently rejected by the Army Nurse Corps, or, as an alternative, to design uniforms using excess WAC material of Army green. But when the day came to display the alternatives to General Arnold for his selection there were three uniforms. The Nurse Corps uniform and another made of WAC material were worn by two Pentagon typists. Beside them, in General Arnold's Pentagon office, a professional model stood in a Santiago blue wool serge belted jacket and A-line skirt, over a crisp white shirt. General Arnold shot a twinkling glance at Jacqueline Cochran and made the appropriate choice.

As illustration of just how military the WASP was becoming, the arrival of uniforms in April 1944, turned into a typical Army snafu. The winter uniforms were of wool, but over a third of the WASPs were stationed at AAF bases in the South where temperatures were already soaring past seventy. By June, orders for lighter summer-weight uniforms could not be filled because all sizes smaller than fourteen were gone. The WASPs had no uniform allowance (AAF second lieutenants got $250). Although the AAF provided two of the basic jacket, skirt, slacks and flying suits, which cost approximately $175, each WASP had to spend up to $100 of her salary on shirts, undergarments and shoes, the latter necessitating the use of precious civilian ration shoe stamps.

Nevertheless, the most exciting fashion shows of the year were in the WASP quarters. Suddenly they looked like officers, as their wings and gold W.A.S.P. insignia flashed against the deep blue uniforms. The WASPs were the first air force fliers in America to wear blue.

While the uniforms were being issued, Jacqueline Cochran sent the first group of fifty WASPs to a four-week course at the AAF School of Advanced Tactics, a type of Officer Candidate School, at Orlando, Florida. If the WASPs had not thought about commissions as AAF officers, they were beginning to now. Classes ran eight hours a day, six days a week. They studied military flow charts, military law and courts-martial. They were shown top-secret bombsights and radar equipment. They were trained in jungle survival, which included how to cook and eat rattlesnake meat, and how to distinguish poisonous roots and ferns. They were given whiffs of mustard gas and cyanide in a course in chemical warfare. Not surprisingly, the old problem of facilities for women arose at Orlando. At class break, the WASPs had to corral their courage, yell into the men's room, "Is anyone there?" and then line up outside to wait their turns, which caused a commotion among on-lookers. On Sundays, relaxing at one of Orlando's lakes, the WASPs found themselves hobnobbing with colonels who were there for other advanced AAF courses, and at the officers' club there was usually a general or two to dance with to an AAF Symphonette. In their new uniforms, the 460 WASPs who would attend officers' training at Orlando got their first taste of what it was like to be a real officer in wartime America. They had come a long way in a year and a half from zoot suits, primary trainers and the dusty plains of Texas.

Finally, to the impressive exterior of legitimacy was added the AAF's official recognition of the WASPs' contribution to its domestic flying operations. As a special part of the graduation ceremonies for the class of 44–2 at Avenger Field, WASP Barbara Erickson, flown into Sweetwater from her ferrying base at Long Beach, was awarded the prestigious Air Medal. Nancy Love, the WASP executive in the ferrying division, on a rare visit to Avenger Field, beamed from the platform as Jacqueline Cochran commended WASP Erickson for a prodigious feat of flying. In only five days, she had delivered a P–51, two P–47s and a C–47 DC–3 cargo transport to their destinations, for a combined 8,000 miles and approximately forty flying hours, as many as most ferry pilots flew in an entire month.

To present the Air Medal, General Arnold made his first trip to Avenger Field. Sharing the platform with him to pay tribute to WASP Erickson that day were all of the WASPs' commanding generals: General Harold George of the Air Transport Command; General Barton Yount of the Training Command; Jacqueline Cochran's chief at the Pentagon, Major General H. A. Craig, Assistant Chief of Air Staff, Operations, Commitments and Recruitment; Brigadier General Isaiah Davies of the Central Flying Training Command, having jurisdiction over Avenger Field and two other generals representing the combined Chiefs of Staff and Training Command Flight Surgeon's office. They all smiled, watching General Arnold and WASP Erickson battle for the brightest blush as the General's strong hands trembled in his attempt to pin the Air Medal over her breast with decorum. And every young WASP in the audience felt a stab of pride that such an honor had been bestowed on one of *them.* (Erickson was the first civilian to receive the Air Medal since Amelia Earhart.)

The next day, Barbara took her Air Medal back to Long Beach, where she was squadron commander. She showed it around to her WASP family like a good report card, then packed it away in a drawer. Curiously, as word circulated around WASP squadrons across the country, many were embarrassed about the Air Medal. They were so indoctrinated that their flying duty was of a lower order than the male pilots in the AAF that many felt it should have gone to an Air Transport Pilot flying the Hump in China, or ferrying bombers across to England. A few saw it as a publicity stunt by WASP headquarters in preparation for the WASP militarization efforts in Congress.

Even Barbara herself thought that any of the WASPs could have done what she did. But that was just the point. Though the Air Medal was given to Barbara Erickson, it was intended as thanks from General Arnold to all the WASPs.

THERE WAS no escaping romance. As there was no mixing with enlisted personnel in the military, air bases with from several hundred to many thousands of men had no more than a few dozen WASPs with whom a lieutenant or major could socialize. Flying officers were among the AAF's most carefully selected young men and women in the country. They had a lot in common—uniforms, flying and sometimes chemistry. And they were in the best physical shape ever.

Far removed from the stern gazes of their hometown neighbors and preachers, and consistent with the extraordinary adventurousness of learning to fly and joining the WASP in the first place, the WASPs often had a more uninhibited attitude toward relationships than many Americans. A good many believed that making love with the man one planned to marry was completely legitimate. The WASP's "Mile High Club"—made up of those who had made love above 5,000 feet—had an enthusiastic, if indeterminate number of members. Once in airplanes where someone else could take over the controls, the WASPs had ample opportunities for consummating a relationship, some of which were spectacular—like in the panoramic plexiglass nose of a B–17 bomber high above the aquamarine waters off the Florida Keys.

There was an irresistible attitude of "Eat, drink and be merry, for tomorrow we fly." Whether they were ferry pilots flying from hotel room to hotel room across the country, or utility pilots stationed at one base, pilots were lucky when it came to socializing in the evenings. If one too many the night before made them unsteady for their early-morning takeoff, all they had to do was nuzzle into the oxygen mask in the cockpit for a minute or two to be back in top form.

The social antics legendary among men on military bases were immediately more sensational when WASPs were involved. At a lawn party given by the base commander of a Texas air base, a WASP arrived perched on the handlebars of her beau's bicycle, with a bottle of beer in her hand. The scene raised many an eyebrow.

The WASPs protected the more spirited of their number. Among the unofficial duties of one WASP squadron commander was to arrange for an abortion. In general, however, behavior was limited by discretion. At an air base in Texas where WASPs were training navigators, off the pilots' lounge was a small room with three beds in it. Ostensibly to be used when pilots were fatigued from long cross-country trips in their twin-engine navigation ships, the WASPs were aware that there were other possibilities for that room. But, conspicuous wherever they were, they realized that with one slip-up, they would all be tainted. In fact, one WASP ferry pilot was threatened with court-martial when base command learned of a liaison with a married B–25 pilot, throwing the WASP squadron into a tailspin, while the male officer remained uncharged.

Nevertheless, the WASPs adopted a popular song of the era, which they sang at Sweetwater with gusto, a sultry ballad called "Rugged but Right" (which even appeared in the WASP Song Book):

I just called up to tell you that I'm rugged but right!
A rambling woman, a gambling woman, drunk every night.
A porterhouse steak three times a day for my board,
That's more than any decent gal can afford!
I've got a big electric fan to keep me cool while I eat,
A tall handsome man to keep me warm while I sleep.
I'm a rambling woman, a gambling woman and BOY am I tight!
I just called up to tell you that I'm rugged but right!
HO-HO-HO–Rugged but right!

ROMANTIC ADVENTURES were but brief interludes in the WASPs' daily flying missions. By June 1944, twenty-three woman pilots had crashed to their deaths. In fact, during 1944, a WASP would be killed almost every month. Most died because of mechanical failures of their aircraft. One was hit midair by another plane when a negligent tower operator cleared two pursuits for landing. Another WASP died when acting as a copilot, with a male pilot at the controls. Several more died at Sweetwater with their flight instructors sitting in the cockpits with them. No matter how good a pilot she was, with seven million people in uniform and many more as civilians attached to the AAF, accidents were bound to happen. War was, after all, a team effort, and the first string was usually in combat. For every pilot killed under enemy fire, four died in "routine" noncombat accidents on the home front.

Training schedules were blitzlike and innovations in aviation lightning fast. After three years of the most concerted research and production effort the world had ever seen, war technology had become so sophisticated as to render its air arsenal almost unrecognizable. As new airplanes were introduced, the AAF's military strategists immediately put them into the action plans. It was here that the WASPs made perhaps their most startling contribution to the war effort. They automatically were put into the cockpits of the AAF's newest and most dangerous airplanes expressly to prove to men pilots that they were so easy to handle that even a woman could fly them.

It has already been recounted how WASPs saved countless lives by developing innovative landing techniques in the P–39 fighter, and how General Tunner sought to bolster morale among Flying Fortress transoceanic ferry pilots by having Nancy Love and Betty Gillies fly the Atlantic. But the AAF Training Command, too, used flight demonstrations by WASPs at carefully chosen air bases to goad men pilots into approaching their training with more enthusiasm. This was first done in the B–26, a twin-engine bomber which when introduced was dubbed the "Flying Prostitute." Its wingspan looked so short, men pilots half-joked, it seemed to have no visible means of support. But for all the kidding, morale among B–26 trainees was low.

In the fall of 1943, an entire training base in Alabama was called unexpectedly to the flightline for a dress review, and was made to watch a spectacular air show by two B–26s. When the bombers taxied up to the formation, four WASPs climbed out of the cockpits. Grumbling among the male pilots was quickly swallowed and the B–26 proved itself by accruing the lowest loss rate of any Allied bomber during the war. Because of this "morale boosting" experiment, over a hundred WASPs trained as B–26 pilots and copilots and performed a variety of flying missions for over a year in the B–26 without a single mishap. (A WASP, however, was killed in a B–26 with a male pilot at the controls.)

As newly designed airplanes poured off the assembly lines, the men who had to take them to war understandably looked at them askance. A sure-fire way to foster enthusiasm, as well as increase their confidence, was to jolt their masculine egos. With this "morale booster" motive, two WASPs were given flight training in the biggest, most costly and complicated airplane ever made—the colossal B–29 Superfortress. In early July 1944, in a major Pacific battle, the Americans had captured the island of Saipan north of Guam in the Marianas, which was only 1,500 miles from the Japanese mainland. By November, a huge air base would be built from which B–29s, with their range of over 3,500 miles without refueling, would take off on continuous bombing raids bound for Tokyo. (Later, raids would originate in Okinawa.)

Meanwhile, back home, moving pilots and crews from B–17s and B–24s to the AAF's newest, most formidable four-engine bomber was a top priority in the AAF Training Command. But proponents of the B–29, among them twenty-five-year-old Lieutenant Colonel Paul W. Tibbets, were encountering extreme reluctance among pilots in willingness to train on the bomber that would ultimately win the war in the Pacific. The Superfortress was twenty-five feet longer, its wingspan was forty feet wider, and its bulk twice as heavy as the B–17 Flying Fortress, now the war-weary work horse of the air in Europe. But pilots were dubious of its flying ability for reasons other than its preposterous size. Thrown into full production after lazing for four years on Boeing's drawing boards, the B–29 had not yet gone through the years of operational tests and modifications that had earned for its now diminutive predecessor, the B–17, its reputation as the Big Friend.

The most alarming engineering bug was that the engine cowlings around the four 2,200-horsepower engines were too tight, causing the engines to overheat and even catch fire on the ground before takeoff. At one of the so-called "very-heavy-bomber" training bases, fires were averaging every fifty hours of engine time.

Colonel Tibbets, based at the very-heavy-bomber training school at Birmingham, Alabama, had heard of the women pilots' electric effect on the training effort in the B–26. He began to comb the area for WASPs. Finally, he learned of a WASP squadron at Eglin Field, in nearby northern Florida. Tibbets went to Eglin and returned with twenty-two-year-old Dora Dougherty and Dorothea Johnson, veterans of Camp Davis and now towing targets for pursuit gunnery practice.

WASPs Dougherty and Johnson began three days of intensive training with their young colonel. Tibbets did not tell them of the high instance of engine fires. They were taught to roll the colossal bomber out onto the ramp as soon as the engines were running, and take off without power checks, as if these were standard procedures.

Meanwhile, a new B–29 on the field was being outfitted for demonstration flights. On the huge nose, painters were recreating Walt Disney's WASP mascot, "Fifinella," and blocking in the airplane's new name, the "Lady Bird." Colonel Tibbets was determined that his women demonstration pilots would cause a sensation.

They did. They flew to the very-heavy-bomber base at Almagordo, New Mexico, and for several days, Dora and Dorothea criss-crossed the state carrying pilots, crew chiefs, navigators, gunners and bombardiers, all of whom would soon be bound for the Pacific. The men could not get over how well the two WASP beauties flew the bomber that had the reputation of being such a fiery beast. They flew so well, in fact, that word reached Washington about the young colonel's sensational training tactics, and the Chief of Air Staff, Major General Barney B. Giles, wired that the two WASPs must stop flying the Superfortress immediately. To boost morale was one thing, Giles told Tibbets, but the two girls were "putting the big football players to shame." By then, however, the point had been made. Two days later, WASPs Dougherty and Johnson were back at Eglin Field, Florida. Though they were two of only about a hundred pilots in the world who knew how to fly the Allies' largest instrument of war, they never set foot in a B–29 again. They would long think with nostalgia of their brief affair with the Superfortress, but they felt they had made an important contribution to the war effort. And that was what counted.

On air bases across the country, the eager WASPs grabbed every oportunity, begged, cajoled, and even stole their way ("bootleg transition," they called it) into the cockpits of as many different kinds of airplanes as they could. One WASP's persistence and willingness to take a chance launched her into aviation's next generation.

In January 1944, Jacqueline Cochran sent Ann Baumgartner from Camp Davis on temporary duty to Wright Field, Dayton, Ohio, the equivalent of Houston Space Flight Center today, to flight test new equipment fashioned for women. When she arrived at Wright Field, Ann thought she had stepped into a Jules Verne futurist novel. Flight engineers and test pilots from all the Allied countries were gathered to pore over plans for rockets, jet propulsion systems, highly accurate bombsights run by rudimentary computers, and experimental navigation and airport systems, like the VOR (Very-high-frequency Omni-directional Radio-range) and the three-dimensional ILS (Instrument Landing System) which would be standard across America three decades later. To Ann's fascination, engineers were even talking of going to the moon.

After her missions flight testing women's oxygen masks and flight suits were over, Ann wanted desperately to stay at Wright Field, and not be sent back to the swamps of Camp Davis. Wright Field's commanding officer told her she could stay if she was willing to work in operations offices at flight test stations. If she did her job well, she might eventually do some flying. Though she dreaded the possibility of being grounded for the remainder of the war, Ann decided to take her chances.

Ann Baumgartner's gamble paid off, and in spades. In no time, she was back in the air, checking out on the P–51 and P–47. She soon found herself part of one of the nation's most elite and daredevil teams of test pilots, experimenting with pressurization and oxygen systems, pursuit gunsights and other prototype equipment. Much of their work was stratospheric and among many other things, Ann's tests revealed a prototype gun firing mechanism that froze at frigid high altitudes. For target practice, they used test islands in Lake Erie, often having to fire rounds of ammunition over the heads of obstinant fisherman in the restricted area.

Every day, Ann flight tested supercharger mechanisms, types of fuel, fuel distribution systems, and numerous engine innovations. She became well acquainted with "dead stick" landings after engine failures, as well as with the sudden surges of adrenaline that accompany danger, and which, like a blast of oxygen, made her astonishingly clearheaded. Like her male test pilot colleagues, Ann came to trust herself in a tight spot. In spite of the thrill of participating in the avant-garde of aviation, however, Ann did not entirely trust the engineering marvels. She had watched one dive bomber test pilot plunge to his death, and whenever she taxied a fighter out onto the runway, she thought, "Well, I wonder if it's me this time. . . ."

But the constant risks were worth taking. Wright Field had some unique planes on the flightline. Ann became one of the very few American pilots to fly a Japanese Zero and a German Messerschmitt, both brazenly stolen by U.S. combat pilots and shipped to America to be studied. But the most spectacular airplane at Wright Field was one that very few of the test pilots had been close to. Ann had heard the eerie scream of its engines floating over the field from a distant flight test station. It was the YP–59, America's first experimental jet.

Even as the WASPs and their male colleagues flew the AAF's biggest bombers and fastest fighters, aviation was in the experimental stages of an entirely new era which would make all of World War II's airplanes obsolete. In the summer of 1944, Allied bomber crews over Germany watched in amazement as a small aircraft shot almost straight up at them, fired a round of .50-caliber bullets, and then disappeared so quickly that if they had blinked, they might have missed it. The craft was a Messerschmitt jet fighter. In June, the horrifying whine of the V1 "Doodlebug" jet-propelled rockets was first heard over southern England as German buzz bombs began their devastating plunges, at approach speeds of over 350 miles an hour. The Allies had few planes that could catch up with one to knock it out of the skies.

The jet was the most profound development in aviation since the Wright brothers made their first heavier-than-air flights in 1903, forty years before. No propeller, no matter how large, and no piston-driven engine, no matter how rapidly it churned, could drive an airplane past 500 miles an hour (versions of the British Spitfire and the Republic P–47 Thunderbolt had attained 465 miles an hour). No longer pulled, a jet was propelled through the air by its exhaust. Merely by increasing the size of its air-compression and gas-combustion chambers, a jet could attain seemingly limitless speeds.

Although the British put a twin-jet Gloster Meteor in action against the buzz bombs, for the Americans the jet was to remain a highly experimental development until after the war. Most American combat pilots would never see one, and many did not even know jets existed. Though none would see combat during World War II, Bell Aircraft had built twelve prototype models of a twin-turbine jet fighter, the YP–59. Several were sent to Wright Field for top-secret flight tests, and in the fall of 1944, nine months after she had risked being grounded by staying at Wright Field, WASP Ann Baumgartner, a twenty-one-year-old former New York journalist, became one of America's first jet pilots, extending women's participation in wartime aviation to its farthest experimental reaches.

As she waited for her first flight in the YP–59, engineers explained to her that the experimental jet turbine engines had a very slow reaction time. Once she established her landing pattern, she had to land. There were no go-arounds. Being a prototype model, the jet carried only half to three quarters of an hour of fuel on board, so she had to complete her flight tests quickly. Once airborne, the flight characteristics of the jet fighter were similar to a P–51's. But the YP–59 would cruise at fifty miles an hour faster.

Ann walked around the small jet fighter in awe, knowing that she was about to become one of America's first jet pilots. The YP–59 was slightly bigger than propeller-driven fighters, but without the heavy engine it was a full ton-and-a-half lighter than a P–47 Thunderbolt. Nestled almost against the fuselage, suspended from the wings, were the fighter's two General Electric turbine jet engines, each with 1,650 pounds of thrust (the F4 Phantom jet of the 1960s had two 12,500-pound engines). The YP–59 fighter seemed strangely nude without the huge propellers Ann was used to. Instead of a propeller, machine gun barrels pointed directly out of the nose of the jet.

As Ann started the engines, she was surrounded by an unearthly, piercing scream, so different from the thunderous roar of powerful piston-driven engines. She buckled her helmet against the noise, and taxied to the end of the runway. Then she gave the engine full power and the jet started forward. Suddenly she was off the ground and climbing steeply. Ann was mesmerized. The noise drifted away and it was quiet in the cockpit; only a persistent high-pitched whine met her ears. Ann climbed up to 35,000 feet and leveled off. Streaking silently across the sky at 350 miles an hour, she tested the oxygen and pressurization system, and fired a few rounds from the .50-caliber machine guns. Half an hour passed like a minute, and she descended to land, taking great care to set up a precise pattern and approach. All too soon the jet's tricycle landing gear touched down on the runway. As Ann ran excitedly to the operations office to report to flight test engineers, she did not know that she would be the only woman jet pilot in America for almost a decade—until Jacqueline Cochran would break the sound barrier in 1953.

By the spring of 1944, the WASP program had burst into full flower; WASPs were uniformed, honored with the AAF's highest fliers' service medal, in training as officers and flying missions with confidence and efficiency on over seventy air bases across the country. But the air was beginnnig to change.

Combat pilots, having flown their fifty missions over Europe, were coming home, some for a rest and then further flight training, but some for good. The Army Air Forces had been training some 100,000 pilots a year, to meet combat losses in the air projected to be as high as 20 percent. But in aerial fighting over Africa in 1943, losses had been much lower—only 7.5 percent. For the first few months of air raids over Europe, losses were under 13 percent. As the spring of 1944 progressed, combat returnees saw to their astonishment that while they had been fighting abroad, a revolution had occurred on the home front.

In May 1944, WASP Virginia Streeter walked out to the flightline at Enid, Oklahoma, to flight test a BT–13 with a rebuilt engine. A young officer was standing by the plane. "What are *you* doing?" he demanded. "Are you tuning this up? Are you a mechanic?"

"No, I'm going to fly it," Virginia answered.

"*You?*" he retorted. "Since when has that happened?"

"I've been here several months," Virginia said, hopping into the cockpit.

Many combat pilots encountered their worst missions and airplanes *after* they were safely back home. Now a seasoned test pilot at Gardner Field, California, WASP Gene Shaffer reported to the maintenance office one morning in June 1944, to be introduced to a lieutenant recently returned from England and assigned to Gardner as a test pilot. Gene was to take him up on her first test flight of the day to show him the testing procedures, in a BT that one of the cadets had mistakenly taxied into another plane. It had a new propeller and a rebuilt wing and elevator. "I've flown fifty missions," the lieutenant told Gene as they walked out to their airplane. "I'm just here waiting for a spot in P–38 school, so I can keep up my flight pay. After all I've been through, now they've got me going up in an old beat-up BT!" Gene sensed that the combat veteran was insulted by his new assignment. "It certainly must be dull compared to the real war," she said, properly respectful.

Gene and her copilot took off, and as Gene put the BT through the required slow-flight maneuvers and stalls, the male voice from the back seat sounded increasingly bored. "Now we do a three-turn spin," Gene announced as she began to climb to 10,000 feet. That perked up her passenger.

"You're going to spin this crate?" the lieutenant gasped.

"Standard procedure," Gene said cheerily, and kicked the rudder pedal. The BT fell over into a spin, and after three spirals she popped the stick forward. But instead of pulling out level, the plane merely changed its angle to the ground and began circling dizzily toward the valley floor like water swirling down the drain. The elevator had not been fixed. "We've got a flat spin!" Gene yelled. "Can you recover better?" The lieutenant tried, but to no avail. Soon both were frantically working every movable lever in the cockpit as the ground came closer and closer. The attitude of the plane was such that the force of the air had locked every control surface into position. Nothing moved. Gene knew there was only one thing to do. "I'm gunning it," she shouted.

"You're crazy!" the lieutenant shrieked.

To gun the engine as they were already plummeting toward the ground at breakneck speed did seem drastic, but as Gene pushed the throttle to full power, the jolt was just enough to disturb the airlock around the ailerons and elevator. The BT shuddered violently, dove briefly and pulled out of the spin. Gene looked at the altimeter. They had fallen to under 1,000 feet.

Neither Gene nor her passenger spoke as they flew back to Gardner Field. But as soon as they threw back the canopy at the hangar, the lieutenant emerged from the cockpit pale as a ghost and angry. "I'm not being paid to do this!" he said, almost in tears.

"Neither am I," Gene said, fanning herself with her clipboard.

Gene never saw the lieutenant again. Flight testing those rotten airplanes, the combat veteran had told the chief maintenance officer, was too damn dangerous. Once again, the Gardner Field maintenance unit squadron was an all-female operation.

At the very moment Gene Shaffer's combat veteran copilot was storming away from the Gardner Field maintenance unit, however, there were thousands of men pilots clamoring for the WASPs' high-risk, no-glory flying jobs. With its combat losses far fewer than expected, it was with gratitude for a job well done that in January 1944 General Arnold closed all primary, or first-stage, flight training schools across the country. (As was the case at Avenger Field, AAF primary training was contracted out to civilian aviation companies. Unlike the practice at Avenger Field, successful male cadet primary graduates went on to basic and advanced schools run by the Army with Army instructors.) At the same time, General Arnold also terminated the Civil Aeronautics Administration War Training Service (originally the Civilian Pilot Training Program) which was training flight instructors for the civilian-run primary flight schools, and for its own WTS flight schools. Furthermore, 35,000 young AAF officers who were on the waiting list for AAF flight training were transferred to the ground forces. With the invasion of Europe imminent, the Army's greatest need was to expand the infantry. Enough cadets were already in basic and advanced stages of flight training, and enough pilots were still performing domestic flying duties, that General Arnold was confident that the future war in the air could be waged and won.

In February 1944, the AAF presented to Congress its bill to militarize women pilots, as it had its Women's Army Corps the year before. Now that its pilot training effort was winding down, never was the necessity more clear to General Arnold for women pilots to fly domestically so that every male AAF pilot could be released for combat duty overseas. Arnold was impressed with the WASPs' performance and delighted that Cochran had so many qualified applicants that her program could be expanded. So General Arnold would argue to the House Military Affairs Committee during hearings on the WASP militarization bill on March 22, 1944.

But to the consternation of the AAF, the civilian aviation establishment exploded in rage when the lucrative AAF-contract civilian primary flight schools were closed and the War Training Service program deactivated. Up to 10,000 civilian flight instructors, or men in training to be instructors, were losing highly paid flying jobs, or the prospect of them. Worse, however, was the loss of their enlisted reserve status, which kept the Selective Service System at bay. They were now threatened with being drafted into the "walking army"—while the AAF sought to commission women as pilots.

As soon as the WASP bill was introduced, flight instructors began a relentless barrage of telegrams, letters and personal visits to their Congressmen. They enlisted vocal support from the American Legion and the editors of many local newspapers and civilian aviation magazines. How could the AAF even think of commissioning women, they protested, when there were now thousands of male pilots available to take over the WASPs' flying jobs?

Suddenly, what was expected to be a routine stamp of Congressional approval on a successful AAF experiment had mushroomed into one of the AAF Commanding General's biggest battles with Congress. As the Allies marched toward D-Day, General Arnold went to Capitol Hill, only to find the enemy so well fortified that he had to fight for the very life of the WASP program.

XV

The Battle of the WASP

AT TEN-THIRTY on Wednesday morning, March 22, 1944, General Henry Harley Arnold sat in full military regalia at the witness table before the House Committee on Military Affairs. Arnold had appeared before the eighteen-member committee countless times during the twenty-seven months that America had been at war. Military Affairs Committee members were Congress' experts on the Army Air Forces, the privileged initiates with up-to-date facts and figures on the progress of the war, and for two Congresses, the distinguished spokesmen for Public Issue Number One. Grateful, they had served the AAF Commanding General well, as his advocates to the public's elected representatives. What General Arnold needed to conduct the war, they were there to get for him. And they had rarely let him down. This morning, General Arnold had come to Capitol Hill with the committee's next mission—to militarize the Women's Airforce Service Pilots.

On the committee was Representative John Costello from Los Angeles, a forty-one-year-old lawyer and veteran of five Congresses, who had introduced H.R. 4219, a bill authorizing AAF commissions for women pilots on duty and appointments as AAF aviation cadets for women trainees. The bill would also grant to women the same privileges, insurance, hospitalization and death benefits given to male flying officers and cadets in the Army Air Forces. Secretary of War, Henry L. Stimson, had written a letter for insertion in the record, giving his complete approval, on behalf of the Roosevelt administration. For the Military Affairs Committee's hearing on H.R. 4219, General Arnold would be the only witness. From past experience, Congress would not in good conscience defeat a bill that the AAF Commanding General declared was necessary to win the war.

"Gentlemen," General Arnold began, "for some time it has been apparent that there is a serious manpower shortage. We must provide fighting men wherever we can, replacing them by women wherever we can." Arnold presented a glowing picture of what WASPs were currently doing for the AAF. "It is not beyond all reason to expect that some day all of our Air Transport Command ferrying within the United States will be done by women. . . . In the tow-target work there is no reason why they should not replace all the men . . . except in the actual combat theaters. . . . Another reason why we must have legislation to make the WASPs a part of the Army is, at the present time they are not entitled to benefits which should go to them in accordance with the duties they are performing, such as hospitalization, authority to live on Army posts, insurance, and some benefits in case of accidents and deaths which they do not get in their civilian capacity."

The committee listened politely. They knew very little about the Women's Airforce Service Pilots. One committee member asked if WASPs were paid, another if they were being used in combat. They waited dutifully for Arnold to make his statement for the hearing record, and responded appropriately. "From our point of view," said Arnold, "with the present terrific manpower shortage we should use every means we can to put women in where they can replace men. This bill will help to do that but will also make far more effective the employment of the present WASPs that we have in our service."

"In other words, the legislation is an emergency proposition?" asked Committee Chairman Andrew J. May of Kentucky.

"It is an emergency proposition so far as I am concerned," said General Arnold.

This was all the committee needed to hear about the WASPs. But they had another matter to take up with the AAF Commanding General, about which they wanted his guidance. Ever since January, Congressmen from all over the country had been receiving letters and personal visits from representatives of two groups of male civilian fliers demanding commissions—the 900 flight instructors and 5,000 trainees released by the AAF's recent termination of the Civil Aeronautics Administration War Training Service (formerly the Civilian Pilot Training Program), and the 8,000 flight instructors who had lost their jobs when the AAF closed its civilian-contract primary cadet schools. No one knew how many of the 14,000 men were actually protesting, but they were fiercely vocal in their criticism of the AAF. They claimed that the AAF had promised them commissions, but instead had thrown out their programs. As many of the fliers had enlisted in the reserve to keep them draft-deferred pending their commissions, they had suddenly been called for active duty and were being assigned as ground crew, in spite of their flying experience. Yet, the men protested, the AAF continued to utilize women as pilots, and now even had the audacity to ask Congress to commission them. For many members of Congress, the letters from these male civilian fliers were their first introduction to the Women's Airforce Service Pilots, and they were not favorably impressed. General Arnold was aware of the controversy.

"We realize that people say that in taking in women we are depriving men who have had a certain amount of training of their possible just due and right," Arnold told the Military Affairs Committee. "I cannot altogether accept that. Our policy is that any man who has had any flying whatsoever will be given a chance to qualify either as a pilot, a copilot, a bombardier, a gunner or a navigator. If they cannot qualify according to our standards in one of these capacities then we offer them other training in the AAF. We cannot lower our standards because a man has had a few hours in the air."

Arnold presented figures enumerating how many CAA War Training Service men had already been dispersed throughout the AAF. Under one third had been able to qualify for aviation cadet training or AAF flying duties. As for the others, Arnold stood firm. "We must not lower our standards because the job that we have to do in the war theaters is such that we must be able to count on every man being able to do his part in a team. . . . As I have said many times, when 1,000 heavy bombers go over Germany there are 10,000 men who go over Germany . . . and if the various people cannot meet our standards, then we regretfully have to release them to some other duty." The WASPs, on the other hand, met the AAF's highest, most stringent requirements, Arnold reminded the committee—those demanded of combat pilots, the AAF's best men.

The Commanding General was clearly annoyed at the civilian pilots' protests. Earlier in the war, the civilian aviation community was called upon to give the AAF's primary phase training to all of its cadets. The AAF and the Civil Aeronautics Administration called for thousands of flight instructors. Men who signed up with the CAA/WTS believed they would be commissioned at the end of their flight instructor training. But the demands of war had changed. Now, in 1944, so many men were in AAF combat pilot training that, as Arnold told the Military Affairs Committee, the AAF's "air program has about reached its peak." Civilian facilities and personnel were no longer needed for flight training. As for the instructors and men in training to be instructors, the CAA's physical requirements were lower than those for AAF flying officers. When the AAF brought the men into the enlisted reserve, its intention was to protect its flight program from losing civilian instructors to the draft, not as a promise of future commissions. "The AAF never promise anybody anything," Arnold asserted to Military Affairs Committee members. "We were probably indirectly responsible because we certainly announced to the public our need for additional personnel; so, indirectly we are always responsible." The committee sympathized.

At this point, General Arnold told the committee he could speak more freely about the civilian flight instructors in executive session, after the portion of the hearings for public consumption was finished. Sensing the direction Arnold was going, the chairman, after fifty-five minutes of open hearing, closed the committee room doors.

Deduced from subsequent statements by Military Affairs Committee members, what Arnold told them in private session would have been highly unpopular, except with his loyal spokesmen in Congress: for commissions, for flying missions and for official membership in the elite air corps, Arnold preferred the WASPs over the male civilian fliers. The WASPs were flying missions that the men would not be able to do as well. Arnold well knew that even within his top-flight air corps, men pilots had been reticent about performing some of the jobs WASPs had taken over, like towing targets and testing planes fresh out of repair depots. WASPs had hopped quickly into the cockpits of airplanes, like the B–26, that his flying officers had been leary of flying. Not only did the WASPs meet the AAF's highest physical and intellectual standards, they were loyal, dutiful and grateful for any flying job Arnold had allowed them to do.

A superb manager, Arnold knew the value not only of recruitment standards, but of attitude. The men civilian fliers, with their display of dissatisfaction, were not endearing themselves to the AAF Commanding General. He was trying to coordinate manpower to win the war; they were complaining about their fate in the war effort. If the men could qualify to be AAF pilots, Arnold was signing them up as fast as they reported to the AAF recruiting center. But he could not help wondering why those who qualified to serve in combat had instead chosen a safer, noncombat civilian job like flight instructing. He found it hard to imagine any one of them heading down a beach prickling with exploding yak-yak guns manned by nervous, raw recruits.

But most important, Arnold insisted on the highest physical standards for his pilots. He refused to give an airplane to an unqualified man. If he could get pilots in top physical condition—male and female—he had no need to compromise, nor would he, merely out of political pressure.

To Arnold, militarizing the WASP was a simple matter. As for the civilian flight instructors and WTS trainees, they were already being assimilated into AAF jobs for which they were qualified. Arnold was commanding general of a force of almost eight million. The disposition of a small group of 14,000 posed no problems, especially when his greatest need was for air support troops, like bombardiers and gunners, and for infrantrymen. He also needed combat pilots overseas, and each WASP released one for him. Arnold could not cater to the individual desires of a few new male recruits.

From a pinnacle of skepticism only two years before about women pilots' capabilities, General Arnold was now completely won over by the performance of Jacqueline Cochran's female flying corps. Whatever he had allowed them to do, they had done, willingly, and well. He could count on them. The controversy mounted by the civilian men fliers boiled down to a question of standards for manhood. And for the AAF Commanding General, it was the WASPs who measured up.

General Arnold left Capitol Hill on March 22, 1944, assured that the Military Affairs Committee would once again take his case to Congress. The committee agreed completely that the AAF must not lower its standards, and if General Arnold said he needed the WASPs, he should have them. By the end of the day, the committee had released a brief two-page report on the hearing, and reçommended passage of H.R. 4219. The bill went to the Rules Committee to await a place on the House calendar. But General Arnold was aware that if his own loyal committee was concerned about them, the civilian flight instructors and War Training Service flight instructor trainees, though insignificant in numbers to him, were a force to be reckoned with if the WASPs were to be made part of the AAF.

THE POLITICAL CONTROVERSY brewing in Congress around the militarization of the WASP was precisely the kind of opposition General Arnold and Jacqueline Cochran had wanted desperately to avoid. They knew they would face skepticism. All of the women's services bills had. At one point in the debate a year earlier on how to utilize uniformed WAVES, a southern Congressman had protested, "Why, bless you, how do you know that they are not going to spend $200 to dress up a girl and then put her in the kitchen?" Yet, despite opposition earlier in the war, women had now joined the military in force. By 1945, the WAC had a membership of 150,000; the WAVES, approximately 100,000; the SPARS, the Coast Guard Women's auxiliary, 13,000; and there were 23,000 Women Marines. In fact, as the battle raged in the press and in Congress over whether women pilots were needed in the air forces, the Navy announced that every month 1,000 uniformed WAVES were being transferred to Washington, to release Navy officers and enlisted men for overseas or sea duty. Its headquarters staff, the Navy boasted in April, was now half female.

But the WASP was different. The jobs performed by other women's branches of the services had not been wanted by men. (The Women's Auxiliary Army Corps, the nonmilitary predecessor of the militarized WAC, was formed in the spring of 1942 to take over motor transport, telephone operator and clerical and cooking jobs.) Jacqueline Cochran had done her best to show that women pilots were taking over flying jobs that the men of the AAF did not want to do. This fact was well known in the Pentagon, but to the civilian public, the WASPs were *pilots*. Pilots during World War II, like the Aces of World War I, were the nation's elite. The Aces of the Air Corps, Navy and Marines, held the same cachet that the astronauts hold today, and all World War II fliers shared in it. No pilot, whether aviation cadet, fighter pilot, or flight instructor, took being grounded with equanimity. It was a fall from grace. Thus, in the WASPs' flying jobs men saw their salvation, and were asserting their right to have them.

When women pilots were first hired by the Air Transport Command ferrying division in September 1942, the AAF perceived a pilot shortage of emergency proportions. Shortly thereafter, they accepted seventeen year olds for pilot training. Jokes ran that they would next be accepting orangutans. Air Transport Command chief, General Harold L. George, in his proposal for the Women's Auxiliary Ferrying Squadron, explained to General Arnold that there was no authorization for flying officers or flight pay in the legislation setting up the Women's Auxiliary Army Corps, or WAAC. Therefore, he urged General Arnold to hire the WAFS on Civil Service status; the AAF could "go to bat" for them later in Congress. But with the formation of Jacqueline Cochran's training program, and the promise of hundreds of women pilots flying with the AAF, militarization became a subject of some debate within the War Department.

As the AAF already had a women's auxiliary, it was assumed that when the WAAC was militarized as the WAC and brought into the AAF by Congress as of July 1, 1943, the WASP would be brought into the WAC. Militarization of the WAC, however, coincided with Jacqueline Cochran's appointment as Director of Women Pilots, placing her on the Air Staff. As Cochran and General Arnold formulated plans to proliferate duty assignments for women pilots, she knew that being part of the WAC would be a bureaucratic hindrance to her progress. For her WASPs to be accepted operationally throughout the air forces, she needed the direct authority of the AAF Commanding General behind her.

She advised General Arnold to keep the WASP on civilian status, ostensibly until the results of her broader WASP experiment were in, but actually to give her time to persuade the AAF to militarize the WASP on its own. Needless to say, many within the AAF, who thought one women's auxiliary—and one female colonel, the popular WAC chief, Texan Oveta Culp Hobby—was enough, saw a gleam of ambition in Jacqueline Cochran's eye.

Amending WAC legislation would have been as complicated and controversial as militarizing the WASP by itself. Age and physical requirements for recruits were different—WACs had to be twenty-one (to assure maturity in the corps), whereas the WASPs were accepted at eighteen and a half. Like their male counterparts, women fliers during their exhausting training did better younger. Furthermore, WACs could not have children under fourteen, while some of the AAF's most accomplished women pilots were mothers of young children.

But underlying Cochran's argument to General Arnold against a WAC-WASP merger was the feeling that fliers were different from ground personnel. Here she could not have found a more responsive chord in the AAF Commanding General. An Air Corps product from its earliest days as part of the Signal Corps—Arnold had learned to fly in 1912—he had long believed that the air forces should be separate from the Army. (This was not done until 1947.) War in the air was unique strategically; the screening and training of pilots had to be done by a command who understood the special demands of flying and aerial warfare. Therefore, General Arnold supported Cochran's efforts to keep the WASP separate from the WAC. But it took them six months to persuade Arnold's boss, the Army Chief of Staff, George C. Marshall, that they were right.

As their number would be small, no more than 2,500 by Jacqueline Cochran's projections, General Arnold realized that the easiest way to bring women pilots into the AAF without controversy in or outside the Pentagon was to commission them directly, the way an equal number of male civilian pilots had been commissioned as Service

Pilots by the Air Transport Command. In January 1944, Arnold asked the Deputy Chief of Air Staff, General William E. Hall, to look into such a possibility. On January 13, the response came back from the Air Staff personnel office. The Comptroller General of the Army Air Forces had examined the law; and, giving eleven pages of documentation, he had ruled that "The authority in the act of September 22, [1941] to make temporary appointments as officers in the Army of the United States 'from among qualified persons' refers to and contemplates *men* exclusively, and may not be regarded as authority for commissioning women as officers in the Army of the United States."

Now General Arnold had no choice but to go to Congress. On February 19, 1944, H.R. 4219 was introduced to face its fate.

By then, the Army's civilian-run flying program had been shut down for a month, and already the civilian flight instructors had aroused the ire of the civilian aviation community against the heretofore little-known women pilots group flying for the AAF. "The fact remains," stated a February 1 editorial in an aviation magazine, forecasting many to follow, "that they are not as suitable for ferry work as men, and now that men are available there is every reason to use them effectively. We are probably stirring up a hornet's nest to suggest that the women withdraw from ferrying fighter and bomber planes, for the WASPs are well entrenched. But the women themselves might look at their own record . . . and feel motivated by patriotic principles and permit trained men, many of them with families, to take over."

By mid-March, before General Arnold's testimony to the Military Affairs Committee on the WASP bill, enough interest had been stirred up in Congress to prompt the Chairman of the Civil Service Committee, Robert J. Ramspeck, a former deputy U.S. Marshall from Decatur, Georgia, to conduct an investigation of the WASP program. As the WASP was under the Civil Service, public funds were being expended for a program about which Congress knew almost nothing. Inquiries had been pouring into the Civil Service Committee from members of Congress on behalf of their civilian flight instructor constituents demanding an investigation to determine if the WASP was worth the money.

When Civil Service Committee Chairman Ramspeck announced his intentions on March 14, Jacqueline Cochran immediately arranged for a meeting with one of the eight women members of Congress, Winifred Stanley of New York, who was the one female member of the Civil Service Committee. A committee investigation, Cochran explained, could "slow up or hinder the AAF's getting the bill passed to make the WASPs part of the AAF." Cochran was worried. After so carefully monitoring the trickle of publicity on the WASPs (a typical headline read, "Girl Flyers Do Well, Ask No Favors"), she now saw that the dam was about to burst. The torrent of public criticism, the kind she most feared, could wash out the program she had worked so hard to create and legitimize.

Functioning in the rarefied inner sanctum of the Pentagon, where General Arnold's word was law, Cochran had been able to avoid public sentiments about women fliers. She had built a functioning organization, but, other than trying to achieve a geographical balance among her WASP trainees, she had not made the slightest attempt to build public support. Her politics had been confined to the Pentagon. Suddenly, just as her goal was in sight—that of a full-fledged military organization—the WASP was being thrown into the public arena to be ravaged, defenseless, by the angriest of critics—men pilots who wanted her WASPs' AAF flying jobs. Among their sympathizers in Congress was Robert Ramspeck of Georgia. Any investigation by his Civil Service Committee would be uncompromisingly biased.

Winifred Stanley was eager to help. She assured Cochran that the highly respected Military Affairs Committee could get the WASP bill through Congress. Meanwhile, Stanley promised to write to Civil Service Committee Chairman Ramspeck that as the War Department had determined the WASPs to be "militarily sound and necessary," a Civil Service Committee inquiry would not seem warranted. But one letter was not enough to stop the investigation.

Back at the Pentagon, Cochran realized that she had to gamble her whole program on one chip—the prestige of the AAF Commanding General. The tenor of the opposition was such that the merits of her case could only be argued by General Arnold and his authoritative spokesmen in Congress, the Military Affairs Committee.

After Arnold's testimony before the Military Affairs Committee, the War Department issued a press release which attempted vainly to extricate the WASP bill from the grasp of the civilian pilots' protests: "The termination of the CAA-WTS program and the subsequent availability of a number of male pilots in varying states of flying proficiency has attracted attention to the Women's Air Force Service Pilot program. The WASP program to utilize women has been found to be militarily sound and necessary. Men released by the termination of the CAA-WTS program can be used for services for which women are not suited, whereas women pilots are qualified for certain flying duties which they can discharge as competently as men. Women are currently ferrying combat aircraft of virtually all types from factories to points of delivery to combat crews. They also fly tow-target planes, planes for gunnery training by air crew members, and do aerial courier work.

"The foregoing is in keeping with the Army Air Forces' policy to employ both men and women in such manner as to contribute the maximum to the war effort."

This clumsy effort to enlighten the public was accompanied by a press conference with Secretary of War Stimson. Then, realizing the futility of trying to make a cogent rebuttal to the civilian male flier's protests, the War Department halted all publicity on the WASPs. So as not to break the hermetic seal around General Arnold's lone testimony that the WASPs were needed to win the war, Cochran decided that the WASPs themselves should be kept insulated from the controversy brewing in the nation's capital. She ordered them not to fly into airfields around Washington, so the press could not nab them for interviews. She also advised them not to write their Congressmen, as General Arnold would be their best representative. Cochran was afraid that the WASPs would be perceived as pushy. But she also did not want to risk being openly opposed by any of "her girls." Instead, Cochran polled her WASPs across the country and on March 29, she wrote proudly to Rules Committee Chairman Adolphe Sabath of Illinois, who would decide if and when H.R. 4219 would be brought to the House floor for debate, that 94.7 percent of the WASPs were in favor of militarization.* And with that overwhelming mandate, Cochran and General Arnold rested their case.

THOUGH they were unaware of the machinations of their militarization bill in Congress, the WASPs had long anticipated commissions as AAF officers. As early as August 1943, two months after Jacqueline Cochran had taken over as Director of Women Pilots, General Arnold had wired Lorraine Zillner, a future member of the class of 44–2 from Chicago, that it was probable the WASPs would be militarized and if she were not prepared to follow through in that event, she should not report to Sweetwater.

Along with the desire to seize an undreamed of opportunity to fly

* Some of the WASPs were opposed to militarization. Many were those with children, who, knowing the restriction among WACs, feared that militarization would force them to resign. One WASP went so far as to transfer legal guardianship of her children to her parents, to assure that she could keep flying after militarization. A few, generally the wealthier WASPs, preferred the liberty of a civilian status.

airplanes for the war effort, in the heart of almost every WASP entering the gates of Avenger Field throbbed a fighting spirit—to prove that she could be just as good a pilot as any man in the AAF. Combined with their shared patriotism and mania for flying, this determination to show women could match up gave the WASPs a lively and universal esprit de corps. To most of them, militarization meant the AAF's final mark of approval, a sign that their guinea pig days were over.

Many months of daily missions as Army pilots, in all but name, had mellowed their fighting spirit somewhat. Their confidence had grown, and the demands and risks of their flying duties distracted them from the crusade they had felt a part of when they were hundreds strong at Sweetwater and striving toward their silver wings. But they retained the sense of being on trial, as if subjected to an arbitrarily prolonged apprenticeship. As the only civilians among military officers and living lives identical to their male officer colleagues, the exasperated WASPs had to explain their stepchild status constantly. But going deeper, as 1944 progressed and they donned their new uniforms, completed their officers' course at Orlando and took over more and more of the flying missions on their air bases, the WASPs yearned for an official rank, as an acknowledgment that they belonged.

As newspapers, newsreels and magazines reported the Allies' step by step build-up toward a European invasion, everyone on the home front was aware that the war effort was accelerating toward major victories. Apprehension and excitement spread around U.S. air bases, as WASP staff pilots watched class after class of AAF cadets, ground, turret and tank gunners, fighter and bomber pilots, navigators and bombardiers graduate and receive their orders to go overseas to participate in the fighting war. Over eleven million men and women were in military service, and never had the military had such prestige, so much appeal, as in 1944. To be left out, or kept at the periphery of the grand Army Air Forces establishment, was a gnawing frustration for the WASPs, no matter how grateful they were to be pilots in the war effort. Though they never dreamed of demanding them, they could not help anticipating their commissions as much as the men around them yearned for an actual shot at a real enemy target.

THROUGHOUT April and May, however the imprecations against the WASPs within the civilian aviation community grew more vituperative. The press had found a sensational issue, and editorials began to appear not only in the nation's aviation magazines, but in newspapers across the country. While the New York *Times* and *Herald Tribune,* and the Boston *Globe* supported giving legitimate military status to the WASPs, many others, including the Washington *Post, Star, Daily News* and *Times Herald,* and *Time* magazine supported the civilian men pilots in the debate. The WASP program was called everything from a "blunder," to a "fast play" and a "racket." In a typical article entitled "Wanted Female—Impersonators" in the April issue of *Contact* magazine, editors railed against "Jackie Cochran's glamor girls." "How about some of these 35-hour female wonders swapping their flying togs for nurses' uniforms? But that would be downright rub-and-scrub work—no glamor there—and we do mean glamor—we sort of remember the 'Airport Annies' who buzzed around our pre-war airports."

"We don't know what the explanation is," an irate Idaho editor wrote. "Probably it is the sentimental softness of American men in regard to their women. In colleges the smooth, good-looking gals can get A's without a lick of work; and in the armed services it may be that dimples have a devastating effect even on the generals."

We think of the war period as being a united front of public sentiment behind the military and political leaders who were running the war effort. But while everyone wanted victory, by 1944, there were already mumblings about the strides American women had made into the nation's work force. The backlash in public opinion which would strike devastating blows after VJ Day was already manifesting itself in respect to the WASPs—that women should step aside when men were again available for their jobs—in spite of the WASP's support from the top.

Throughout the spring, the WASPs themselves began noticing a distinct change in atmosphere. WASP ferry pilots stationed at Long Beach, California, which was near scores of CAA-contract flying schools which had been closed, encountered groups of flight instructors who were pouring onto the air base to take Air Transport Command physicals and flight tests. The men glared at the WASPs as if they had just whisked away their last meal. "What are you doing here?" one group of men shouted. "We have to be here. You should be at home." The WASPs based at Long Beach were almost all fighter pilots and twin-engine rated. The flight instructors, who had spent their air experience in the back of primary trainers yelling at students, had to show the Air Transport Command they were able to fly bigger airplanes with a minimal amount of transition training.

Flight tests were stiff, and many did not pass. The WASPs were cruel reminders they could not make the grade.

From a carefully guarded, top-secret experiment, the WASPs had suddenly become notorious. Having delivered a P–47 to North Carolina, WASP Jill McCormick was sitting in a hotel lobby in downtown Raleigh, reading a book while she waited for the departure of her commercial flight back to her home base at New Castle, Delaware. Suddenly she was surrounded by men in uniform, shouting at her. "Go back home, WASP, we don't need you! You're in a crummy organization, it shouldn't even exist." Then they called her a slut. Dumbfounded, Jill got up from her chair and tried to escape to the ladies' room, but the men would not let her pass. Suddenly, an officer pushed through the crowd, saying, "Leave her alone! She's doing her part like the rest of us!" The men dispersed, grumbling. Jill was now shaking. The officer sat down next to her and tried to calm her. Shortly the bus to the airport was announced. When Jill thanked the officer and said good-bye, his face fell. "You have to leave so soon?" he asked. "I was hoping you'd join me in my room for a drink." Jill ran for the bus feeling sick to her stomach.

Those WASPs who were aware of the criticism were bewildered and hurt. "I simply won't stand by and let them say we aren't needed or aren't doing a good job," Eileen Roach wrote her family in Phoenix, Arizona, from Liberty Field, Savannah, Georgia, where she was towing targets for ground artillery gunners. "All these guys who are complaining about the WASPs should have to do the jobs them selves for a while and see how they like it. This work is detested and looked down on by regular pilots and they think it's beneath them. From our standpoint nothing's beneath us and nothing's too good for us either."

Though the WASPs were not allowed to contact their Congressmen, some of the WASPs' parents wrote on their behalf. But they were a mere whisper in the roar from the civilian flight instructors, the civilian aviation community, the American Legion and the anti-WASP press. In early May, Jacqueline Cochran flew to Orlando to confer with the Director of Training about the progress of the WASP course, and to speak with her WASPs in officer training, among them the WASP B–17 pilots. Cochran found that there was considerable confusion as to why they were all being sent to Orlando. Some had been interviewed by investigators from the Civil Service Committee (for whom Orlando promised the largest concentration of WASPs) and wondered if their commissions were imminent. But Cochran did not want to talk about the investigation, and seemed subdued about their prospects for militarization. "It's out of my hands," she told B-17 pilot Charlotte Mitchell. "We just have to wait to see what Congress decides."

THE WASP bill faced a peculiarly conservative Congress in 1944. In the middle of his third term, which had begun in 1940, President Roosevelt had not been on the ticket in 1942 to encourage progressive votes, and two large, traditionally liberal constituencies had been largely disenfranchised when Americans went to the polls: labor, which in the vast migration to the nation's war plants, failed to meet residency requirements; and the highly mobile twenty-one to thirty-five year age group, millions of whom were flooding the armed services, where they had no vote.* The result was that in the Seventy-eighth Congress, Democrats had only a thirteen-member majority in the House. And there were a slew of first termers who saw in the civilian pilots' cause a hot reelection issue.

General Arnold's Congressional "tournaments on behalf of the faire Cochran" (in the words of one Washington columnist) and her WASPs were more than just ceremonial jousts. He was able to get an item in the $47-billion military appropriations bill to provide for $6 million for the WASPs. But on May 18, days before he left for Britain to oversee last-minute preparations for D-Day, Arnold had to rush to Capitol Hill to fight for the very life of the WASP program.

The House Civil Service Committee investigators had compiled their report and their conclusions were dire. The WASP experiment, it had decided, had resulted in "wasted money and wasted effort." Recruiting new WASP trainees competed with recruitment efforts of established women's services, to their detriment. In fact, the WASPs had already drawn young women away from vital war jobs. (The committee cited cases of an airplane motor inspector, a lens polisher and a War Department map expert.) The training of women pilots at this time of the war, the Civil Service Committee continued, would add to the "tremendous surpluses of trained and experienced pilots throughout the world" after the war, which will "constitute an acute post-war problem."

As far as the Civil Service Committee was concerned, its most devastating indictment, one which assured the concurrence of many conservative members of Congress whatever their views on female

* See Richard R. Lingeman's *Don't You Know There's a War On?* (Putnam, 1970), a unique, fascinating description of the home front.

pilots, was that Congress had never authorized the program, though to train each WASP the AAF had expended $12,500 of taxpayers' money. "Public funds," the committee wrote, "are made up of the War Stamps of school children, the taxes of the farmer, the savings of the wage earner, deductions from the pay envelope of the laborer, and the earnings of industry." The War Department claimed that it found its authority in the fiscal 1943 military budget section under the aegis of "salaries and wages of civilian employees as may be necessary," and "the training of such civilian employees." The committee did not find the WASP program necessary.

For an hour, General Arnold argued with Civil Service Committee Chairman Ramspeck that the WASP program was vitally necessary to the AAF, while Jacqueline Cochran sat in tense silence by his side. Five of the eighteen committee members, among them Winifred Stanley, did not agree with the majority report and asserted that whether or not the WASP was needed for the war effort was a matter for the War Department to decide. Finally, Arnold was able to reach a compromise. The WASPs already trained, or in training, the committee conceded, should remain on duty. But Chairman Ramspeck stood firm against the recruitment of new WASP trainees. Some provision for hospitalization and insurance would be recommended for the current WASPs. But in return, the AAF was directed "that the services of the thousands of civilian flight instructors and trainees be immediately utilized." Arnold assured the committee that he was giving the men every consideration, and urged them not to use the WASP as a pawn.

Arnold's tournaments were not over. Since March 22, when it had been reported out favorably by the House Military Affairs Committee, H.R. 4219 had been languishing in the House Rules Committee, in charge of putting bills on the House calendar for floor debate. It was now the end of May, the House was scheduled to recess for the summer at the end of June, and still no action had been taken, although the calendar was rapidly filling up with hundreds of measures that needed action before the recess. Rules Committee members, under pressure from supporters of the civilian flight instructors, would not let the WASP bill onto the floor without a concession from the Military Affairs Committee to allow an unrelated amendment to be introduced during the debate on the WASPs, which would commission the male civilian fliers.

General Arnold met again in executive session with the Military Affairs Committee. Arnold was loathe to have the WASP bill formally wed in the debate to the CAA fliers. Nor did he want the two issues voted on as one measure. But he had no choice. He agreed to allow a bill sponsored by Military Affairs Committee member Overton Brooks of Louisiana to be introduced during the debate as an amendment to H.R. 4219. Brooks' amendment provided for commissioning the trainees of the War Training Service, "if they met Army Air Forces standards for flying officers." This was not much of a concession, but, at least, the WASP bill would now come to the floor.

ON JUNE 4, Allied troops entered Rome. On June 6, 250,000 men, in planes, gliders and landing craft, left British shores to hit the Normandy beachhead. The day in between, the House Committee on Civil Service, with great fanfare, issued its report on the Women's Airforce Service Pilots, and it was heard as a call to arms for the upcoming floor debate. "If it is necessary at this stage of the war to embark upon this costly and experimental program, then this Nation, insofar as manpower is concerned, is in a worse position than any of our allies, and apparently any of our enemies," said the report. Opponents of the WASPs rallied around the committee. "The House will face a battle of the sexes," exclaimed the Washington *Post,* "when it considers the bill to militarize the WASPs."

On June 19, as a record-breaking summer heatwave engulfed the nation's capital, Congressman James Morrison, a first-termer from Louisiana and one of the WASP's harshest and most persistent critics in Congress, stood up on the floor of the House of Representatives to present the "real facts concerning the controversy between the WASPs and the CPT-CAA-WTS programs." The WASP "was the most super-duper and glamorous of all programs," he drawled. Their "natty and stylish uniforms" were "tailored on Fifth Avenue in New York and cost over $500." The AAF planned to spend $100,000,000 in the training of 5,000 WASPs, an "elite corps" which would have no enlisted personnel. Only 30 percent of the WASPs wanted to be militarized, Morrison asserted. But "perhaps the irony of the whole WASP program is that many instructors of the WTS program with 2000 hours or more in the air . . . may probably clean the windshield and service the plane for a glamorous WASP who has only 35 hours of actual flying time." And thus the debate on militarizing the WASP began.

The following day, June 20, the House took up House Resolution 564, the rule under which it would debate and vote on the WASP bill. As a Rules Committee member presented the resolution and explained provisions of the WASP bill and its Military Affairs Committee amendment to the House, he was interrupted so often by civilian pilot supporters that it took forty-five minutes to bring the rule to a vote. Finally the rule was adopted and the House adjourned until Wednesday, when it would launch into a four and a half hour debate on the fate of the WASPs.

By ten-thirty the next morning the House debate on H.R. 4219 was in full swing. The members of the Military Affairs Committee began asserting their position. "Both General Arnold and the Secretary of War, whose letter appears in the report, stressed the point that the primary purpose of the legislation is to release male pilots for combat training and assignment," said Military Affairs Committee member Charles Elston of Ohio. "This reason appears to have been overlooked almost entirely by the Civil Service Committee and by those who are seeking to be commissioned regardless of their qualifications. . . . Instructors are required to pass only the Class 2 examination, which is the equivalent of the airline pilot test. On the other hand, a WASP must pass the combat examination. . . . The women who are today piloting all kinds of planes across the country are rendering a magnificent service to the country. They should be accorded all of the benefits of a soldier and are entitled to be commissioned as provided for in this bill."

"I am not objecting to them taking the WASPs into the Army, I am not objecting to them utilizing the services of women fliers, but I do believe in simple justice and in economy," protested Congressman Forest A. Harness of Indiana, who had introduced a bill to commission men civilian fliers whatever their qualifications. "Some men were rejected as combat pilots because there were one or two digits of a finger missing or because of flat feet . . . It is this group that we are trying to keep in the service, which can do the job of flying military airplanes that the WASPs are now doing."

When WASP bill sponsor John Costello rose to defend his bill he found himself in a duel with Congressman Compton White of Idaho. Characteristic of the debate:

> White: What is it that these women are qualified to do that these C.A.A. pilots cannot do?
>
> Costello: The C.A.A. pilots can qualify, probably for many of the same jobs, but what the Army needs now is fighting men.
>
> White: And the gentleman wants to take these men out of the flying corps and put them on the ground?
>
> Costello: No.
>
> White: That is the meat of the coconut, is it not?

> Costello: No. If the men are qualified to fly planes, we want to put them in the Army flying planes. If they cannot qualify to fly planes we want to put them in as Army navigators and bombardiers. We want to use every man that is qualified.
>
> White: The gentleman does not contend that these women can qualify for anything that these C.A.A. pilots cannot do?
>
> Costello: I simply state that these women are not going to displace any men. . . . Let me call your attention to this one fact. The sole purpose of this bill is simply this . . . to take these women . . . and convert them into a military capacity. . . . This should be done, because these women at present are denied hospitalization; they are denied insurance benefits, and things of that kind to which, as military personnel, they should be entitled, and because of the work they are doing they should be receiving. . . . The cost of training one of these women is no different from the cost of training a man. . . . The cost for uniforms is the same. . . . The casualty rate is approximately the same. There has been no difference whatsoever between the men and the women."

But Costello was like Isaiah crying in the wilderness. The debate was not on how good the WASPs were as pilots. "If as has been suggested, it is an experiment and if it was successful," grumbled Congressman Edward Rees of Kansas, the ranking minority member of the Civil Service Committee, "why in the world did they embark on a program of spending hundreds of thousands of dollars to do this sort of thing? It seems to me the experiment was carried pretty far."

"I was out of sympathy entirely with the idea of having the WACs and the WAVEs to begin with," said Military Affairs Committee Chairman Andrew J. May of Kentucky. "Perhaps I have some old-fashioned notions about that. But when such men as General Marshall, General Arnold, and the Secretary of War have come to me with the statement that it is essential to the prosecution of this war that they have a certain piece of legislation, I have been pretty consistent in following that advice. I think that is the kind of advice the House of Representatives should follow."

The debate elicited a long speech from Congressman Karl Stefan of Nebraska, now a familiar part of debates on just about anything in the House of Representatives, on the contribution to the war of American womanhood. "No matter what this House feels about the women in our armed forces, Mr. Chairman, I feel now that we are discussing them I cannot resist in some way championing their cause.

My information is voluminous regarding the ability of these women in flying these monsters of the air through storms and clouds and making safe delivery after thousands of miles of flight. The knowledge of some of these women regarding the reading of maps and the handling of radio and their skill in emergencies are contained in many chapters of thrilling experiences of the Army Air Corps. It will be told more graphically when the war is over. That women are rendering outstanding service should not be denied, but it should be acknowledged again at this time when we are dealing with legislation affecting their future. . . . I hope that nothing be done here to eliminate them and that those women flying transports will . . . be given the rewards to which they are entitled. . . . It is not only in aviation that our women are helping to win this war, Mr. Chairman. In the homes of America, Mr. Chairman, the fires of hope are burning because the American women are making that hope live." And so forth.

"I do not believe there is a Member of this House who has gone into this matter as thoroughly as I have," said Congressman James Morrison of Louisiana, who had started the WASP debate two days before. The civilian male pilots "have gotten a raw deal and this WASP program makes it worse." He offered an amendment to limit the WASPs to 1,500.

This brought a resounding slap on the wrist from Military Affairs Committee Chairman May. "We might as well say to the armed forces, 'You can have 1000 tanks and no more,' or 'You can have 5000 planes and no more.' It is the kind of limitation that ought not to be voted by the House of Representatives. . . . We have just entered the Continent of Europe. We are only a few miles from the edge of the water and it is many hundreds of miles onto Berlin. I certainly am not in a position myself, and I do not think anybody else is, to say how many of these pilots may be needed. I therefore suggest that the House vote down this amendment." This was May's first test of strength. The amendment was voted down—but only by nine votes.

At this point, Congressman Joseph P. O'Hara from Minnesota rose to set the record straight. The House, he said, was "losing sight of some of the principles involved in this bill. As far as I am concerned the bill is about as unpalatable as it can be, no matter how it is amended. . . . This bill is apparently for the purpose . . . to commission these very charming young ladies who have been doing a very fine job in flying. Some of them are very able and very fine pilots. There is no question about that. But when you get down to the point where we cannot give any consideration to the thousands and thousands and thousands of fine youngsters who have been washed out because some Army pilot instructor said, 'Well, he does not quite have that which is necessary for a combat pilot' . . . that boy must go back and become a grease monkey or tail gunner . . . because his natural flying ability does not fit into this program . . . because they say, 'We have to have somebody in there who is a very attractive lady pilot.' . . . Why this is a piece of social legislation in my opinion and that is all it is."

One of his colleagues agreed. "I think it is time to forget the glamour of this war and think more of the gore of war."

Costello protested. "If you like to be covered with grease, if you like to sweat out piloting an airplane through stormy weather from one coast to another and call that a social activity, very well, then vote against this bill. I am not going to set myself up here and state that because you give a uniform to these women pilots, just as you have given a uniform to the women nurses or to any one of all the other women's organizations, you are creating some glamorous organization."

"It seems," said Congressman William J. Miller of Connecticut, "that the House is in a bargaining mood today. We are saying to the Chief of the Air Corps: 'If you want 5,000 WASPs, we will let you have them, but if you get them you have to take this other group of students and instructors that you say you do not want at this time.' "

So far during the debate, only two of the three sections of the WASP bill had yet been read on the House floor by the clerk. But Representative Edward Izak of California had had enough. He moved to strike out the enacting clause of H.R. 4219, a parliamentary trick to bring a bill quickly to a vote. To vote to strike the words "Be it enacted . . ." from a bill, whatever its content, killed it with one swift chop. "My object here is to kill this bill . . . because we do not need such a bill." Izak shouted as Military Affairs Committee members stared in disbelief. "It is the most unjustified piece of legislation that could be brought before the House at this late date. I know that any woman would like to have 2,500 girls under her and be a colonel. . . . There are more than 2,500 men sitting out on the beaches of California today, who have been instructing for four years, the finest aviators we have in this country. The Army says, 'You cannot pass the examination, so out you go, but we will uniform these women and let them take your places.' Is that not a fine situation?

. . . I am sorry to see Hap Arnold lose his balance over this proposition . . . I think this is an unconscionable bill."

Military Affairs Committee member R. Ewing Thomason of Texas quickly rose in an attempt to restore the committee's case. "I think I sense the temper of this Committee [of the whole House] and I am taking these five minutes to beg and plead for clear straight thinking without any bias and without any prejudice. It is now 2:30 in this country and the latest press report states that our boys are marching into Cherbourg. The latest press reports also say that a mighty naval battle is now raging in the Pacific. [The Battle of the Phillipine Sea, known as the "Marianas Turkey Shoot," in which the Japanese fleet was badly damaged.] We have got to trust somebody in this terrible war . . . and somehow or other I feel that General Marshall, General Arnold and Admiral King know more about this situation than we do. . . . The record I hold in my hand is the testimony of General Arnold before the Committee on Military Affairs and he testified these young women have done a magnificent job . . . I want to keep up my 100 percent war record."

By this time the debate was such a muddle of confusion, and the House was so exasperated, that neither issue was getting a clear airing. Certainly the merits of the original bill, to militarize the WASPs, had all but been buried in rhetoric either about the civilian fliers' mistreatment or about the latitude Congress should allow the Commanding General of the Army Air Forces. As Congress faced many more bills on its calendar before the summer recess only nine days away, patience was in short supply on the floor of the House. But now that it was faced with a vote to strike the enacting clause of H.R. 4219, after almost five hours of debate, the decision was at last a clear question of whether or not to commission women pilots in the Army Air Forces. Military Affairs Committee Chairman May called for the roll. Committee members listened tensely as their colleagues in Congress shouted their yeas and nays. House members with familiar names in our current world all voted against killing the WASP bill: Everett Dirksen, Christian Herter, Lyndon Baines Johnson, Estes Kefauver, Clare Booth Luce, Mike Mansfield, Hugh Scott and Margaret Chase Smith. But when the role had been called, the final vote was 169 against, 188 for killing H.R. 4219. The WASP bill, its enacting clause stricken, had been parliamentarily castrated.

XVI

Fifinella's *Last Flight*

AFTER FOUR MONTHS of waiting and worrying, Jacqueline Cochran realized that decisions had to be made and made fast. Her militarization bill had been defeated by only nineteen votes. That left her current WASPs at status quo. But, judging from the Civil Service Committee report, ample statements made on the House floor during the debate on H.R. 4219, and the ultimate defeat itself, it was obviously the will of Congress that no more WASPs should be trained. Of most immediate concern were the 113 members of the class of 45–1 who had orders to report to Sweetwater on June 30.

By Monday, June 26, General Arnold was back from England and in his office at the Pentagon. He agreed with his Director of Women Pilots: the class of 45–1 and the 3,000 other young women (screened from over 25,000 applicants) who were accepted and awaiting class assignments would never become WASP trainees. On Monday morning, telegrams went out notifying 45–1 that, following Congressional directive, the WASP training program had been terminated. Over a third of the class, however, had already begun their long train journeys to west Texas. Cochran telephoned Leoti Deaton, her WASP staff executive in charge of Avenger Field. "Deaty" would have the unhappy task of sending 45–1 back home again. To ease the blow slightly, the AAF offered free air transportation for their return trips across the country by the Air Transport Command. It was the first time any women pilots had not had to pay their own way both to and from Avenger Field. It was the least the AAF could do.

On Monday, June 26, General Arnold also appointed a board on his Air Staff to recommend what the AAF should do with its women fliers. Cochran immediately went to work. The War Department legislative liaison office surveyed Congressional staffs and found that as long as the AAF was diligent in its efforts to assimilate the protesting male civilian fliers, there was still support for militarizing the WASPs who were already in training and on duty. Wires and letters had been pouring into Congressional offices from the families and friends of 45–1 and the other 3,000 women pilots accepted by the WASP program and now denied entry to Sweetwater. Congressmen from Texas, including House speaker Sam Rayburn, were interested in keeping Avenger Field operating. But Cochran also learned that the Civil Service Committee was already threatening to introduce legislation, as soon as Congress returned from its summer recess, to freeze the WASPs' civilian status to place greater pressure on the AAF to take in the CAA men as flying officers.

Having placed her trust in General Arnold's prestige and the power of the House Military Affairs Committee, Cochran now decided it was time for her to go on the offensive. She had little to lose. On August 1, she presented General Arnold with an eleven-page, legal-size, single-spaced report which gave a complete history of the WASP program, an enumeration of all ratings and flying assignments for her 773 WASP graduates, and her recommendations for the WASP's future. The War Department ban on WASP publicity was lifted, and on August 7, Cochran's entire report was released to the press. For the first time since the WASP controversy began, the WASP program was set before the public on its merits—without reference to the male pilots in Civil Aeronautics programs other than to mention that they lobbied against the WASP bill. Instead, Cochran wrote in introduction, "It is timely to evaluate the service of the WASPs against a background of two years of accomplishment and to determine their future in the light of today's known factors" about the war, and the current needs within the AAF. A detailed response to the Civil Service Committee report, Cochran's exhaustive memorandum should have been in every Congressional office and on every editor's desk in June. It might have changed nineteen minds.

Now Cochran had these alternatives: de-activate the WASPs immediately; keep the WASP as it was; keep its Civil Service status, but attempt legislation to provide hospitalization and insurance, as recommended by the Civil Service Committee; or continue to press for militarization. The last option was perhaps the most hopeless, yet for Cochran, had the most honor and dignity. If the WASP was never to gain the legitimacy of military status, as had all of the other women's services, it would not be because their commandant backed down.

Interspersed with the facts in Cochran's report were fighting words from the Director of Women Pilots, not only against the Civil Service Committee's observations about her program, but beyond. Cochran was finally trying to call forth public support for the WASPs who had been wronged by their government. With the termination of the WASP training program, Cochran wrote, the "hopes of thousands who believed women have the same right as men to serve with their best ability were dissipated." The WASPs' role in the war effort had been totally misconstrued by critics. "The usefulness of WASPs," Cochran asserted, "cannot be measured by the importance of the types of planes they fly, for their job is to do the routine, the dishwashing flying jobs of the AAF, that will release men for higher grades of duty. They are carrying their own freight, doing these jobs while the experience is being gained by the AAF. . . . At present each WASP saves one less-qualified man from being withdrawn from civilian life or releases one already trained pilot for other duties."

After almost two years proving their capabilities and effectiveness as military fliers, Cochran would not tolerate second-class status for her WASPs. The only "means by which the AAF can obtain efficient and economical use of women pilots," Cochran argued, "is through a militarized program which makes the WASP a part of the AAF. Simple justice for the WASPs themselves also dictates such a step. . . . They get none of the benefits of military status, not even the right to a military funeral. As a WASP's pay is less than the income of a second lieutenant on flying duty, it follows that the AAF is getting results at less cost. . . . They are entitled to equal pay and equal recognition for equal work."

In conclusion, Cochran gave the AAF Commanding General a brazen ultimatum: the WASPs should be commissioned into the Air Corps. Moreover, "serious consideration should be given to inactivation of the WASP program if militarization is not soon authorized." Cochran pleaded that if "such action should be taken, an effort should be made to obtain military status, if only for one day, and resulting veterans recognition for all who have served commendably."

On June 22, 1944, the day after the WASP bill was defeated in Congress, President Roosevelt signed into law the omnibus veterans legislation known as the GI Bill of Rights. If the AAF's women fliers were not to enjoy their rightful ranks, Cochran demanded that, at the very least, they receive the government's gratitude for serving and dying for their country.

Cochran's report, the first ever of such magnitude on the WASPs, fascinated the nation's press, and long articles ran across the country detailing at last what the women pilots were doing in the war effort.

But passions had already been raised against the idea of women flying military airplanes, and editorials again appeared against "that WASP bill." Columnist Drew Pearson (who would be voted in October 1944 the most influential columnist in Washington) joined the voices of protest, in an August 6 article called "There is Still Some Sting Left in the WASPS." Pearson was irate at "Arnold's efforts to side-track the law by continuing to use the WASPs while more than 5000 trained men pilots, each with an average of 1250 flying hours remain idle," and "hundreds of Air Corps pilots retiring from combat are anxious to stay in the Army as transport ferry pilots. . . . Magnetic Miss Cochran seems to have quite a drag with the 'brass hats.' "

The civilian flight instructors and CAA War Training Service instructor trainees had yet to get satisfaction from Congress (they never would) and they continued their efforts to spread opposition to the WASP. For example, the Wisconsin Department of the American Legion, at their annual convention early in August, overwhelmingly passed a resolution on their behalf, recommending "the immediate and honorable termination" of the WASPs, "in the interest of preventing excessive waste of available skilled manpower and the squandering of millions of dollars of the tax payers money." A women's flying corps was perceived as a frivolous program, and there was no conception of the money wasted should a thousand government-trained women pilots be sent home.

Thus, as Cochran faced her public opposition head-on, she was fighting a losing battle. But far more serious, she was about to face the defection of her commanding officer. General Arnold had valiantly expended a sizeable amount of political clout in his lone advocacy of militarizing the WASP. Both he and Cochran had underestimated the threat posed by the WASPs, and the power that a small group of men could wield, even against the AAF Commanding General and the Secretary of War, when it came to the issue of legitimizing women's place in the cockpit. After the Comptroller General's ruling that the "qualified persons" whom Arnold had authority to commission did not include women, any further militarization efforts on behalf of the WASPs would require legislation. Arnold was loathe to return to Congress, and for good reason.

Furthermore, in August 1944, the news from the battlefronts was exuberant. The Allies were rapidly securing the entire northern coast of France and would soon invade the French Riviera. On August 25, they would make their triumphant march into Paris. In Italy, troops prepared to attack the formidable but final German fortifications along the Gothic Line in the mountains of Tuscany. And in the Pacific, the Marines were mopping up the Marianas, from which B-29s, so far forced to operate out of air bases in China which were almost impossible to supply, would now have an easy 1,500-mile trip to mainland Japan. General Arnold was not about to face certain defeat on the home front. The women pilots idea had come—and gone. Jacqueline Cochran offered him her ultimatum on the WASP, and publicly, she was in effect giving the AAF Commanding General an honorable way out of the controversy. He soon took it.

On Tuesday, October 3, WASPs now stationed on almost ninety air bases across America returned to their barracks after the day's flying missions to find envelopes addressed to them from AAF Headquarters, Washington, D.C. Some hoped that an official communication from Washington meant that their AAF commissions had at last been approved. But it could also mean something much worse. Those envelopes were opened gingerly. Inside were two mimeographed letters. The first, they saw, was from Jacqueline Cochran. With the first line their hopes were dashed. "To All WASP: General Arnold has directed that the WASP program be deactivated on 20 December 1944. Attached is a letter from him to each of you and it explains the circumstances leading up to his decision." They quickly glanced at Arnold's letter: "I am very proud of you women. . . . So I have directed that the WASP program. . . . I am sorry. . . ." So it was true. Their military flying careers were over.

As if Cochran knew the state of shock into which her readers would fall, her letter continued in an emotionless enumeration of details. She described types of discharges, the Civil Aeronautics Administration's willingness to issue civilian ratings on a par with their military flying experience, requirements for final discharge physicals and logbook certifications. "Those who wish to continue to fly for the Army Air Forces will be disappointed," Cochran wrote. "But no WASP familiar with the pertinent facts and trends would question the decision or its timeliness." In closing, Cochran assured her WASPs, "Each of you has made an important contribution to your country at war and has aided immeasurably in establishing women's place in aviation." To a good many of her WASP readers, Cochran's words swam in a blur of tears.

General Arnold's explanation of his decision was simple. "When we needed you, you came through and have served most commendably under very difficult circumstances," Arnold wrote, "but now the war situation has changed and the time has come when your volunteered services are no longer needed. The situation is that if you continue in service, you will be replacing instead of releasing our young men. I know that the WASP wouldn't want that.

"I want you to know that I appreciate your war service and that the AAF will miss you. I also know that you will join us in being thankful that our combat losses have proved to be much lower than anticipated, even though it means the inactivation of the WASP. . . . My sincere thanks and Happy Landings always."

The disbanding of the WASP, the first of the women's services to be released, was presented to the nation by the War Department as a triumphant sign that the war was being won. The war situation had indeed changed from that of September 1942, when, as Jacqueline Cochran described it in her August 1 report, "the United Nations were marshalling their every resource and praying for time." The Allies were losing. Now two years later, they were winning. The Russians were about to enter Norway. France was liberated, and Allied troops were within twenty-five miles of the German border. In a matter of days, General Douglas MacArthur would make his famous return to the Philippines, and the entire Japanese Navy would be defeated at the Battle of Leyte Gulf. "The decision to disband the WASP," the War Department announced in its press release, "was based on indications that by mid-December there would be sufficient male pilots available to fill all flying assignments in the U.S. and overseas, thus cancelling the need for women to fly the routine jobs in this country. The WASP deactivation is the first during this war of a sizeable group of women serving directly with the armed forces. Such deactivation actually is well underway." Then the release quoted the AAF's grateful Commanding General: " 'I am proud of the WASPs and their record of skill, versatility and loyalty. They have done outstanding work, even exceeding expectations. What they have proven, including flying B-29s, would be of inestimable value should another national emergency arise.' "

To a nation that had lived so long in daily fear of that fatal telegram that a loved one was dead, the news that women were no longer needed meant that their menfolk were coming home.

THE WASPs *were* glad the war was being won. They tried to feel proud that they had contributed to the war effort, and to women's place in aviation. But personally, their lives had just been shattered. To be sent home, long before the war was over (VE Day, May 8, 1945, was many months away) and just when they felt they were contributing of their skills the most, was a devastating and bewildering blow. They were being deprived of "being there" when the glorious moment of victory finally came. But as they walked out to the flightline, after reading Cochran's and Arnold's letters, the pain was immediate and excruciating. Flying had been their lives for over two years. There were no pilots anywhere in the Army Air Forces more avid, enthusiastic and certainly more grateful than the WASPs. They had *loved* their P–51s, B–25s, B–17s. Every second in the air had been precious to them. What were they going to do now?

The specter of December 20 loomed like a thundercloud on the WASPs' horizon. At some bases, cooperative commanding officers let their WASPs use the last few weeks to check out in the larger airplanes on the field, or for the WASPs serving as copilots, in the first pilot's seat, to up-grade their flight logs while they still had access to military airplanes. On other bases, WASPs were barely flying at all, as base commanders scrambled to find and train male AAF pilots to fill the WASPs' places. A joke circulated among WASPs in the Training Command, a conversation between one of the fourteen women test pilots stationed at Pecos, Texas, and one of the ten at Yuma, Arizona, two AAF pilot training bases:

> Pecos WASP: Did you know that they just shipped in fifty new pilots yesterday?
> Yuma WASP: Is that right?
> Pecos WASP: Yes, they brought them in to release the WASPs for *active* duty.

But the prospects for the future of the nation's highly trained female AAF pilots were no laughing matter. In early November, WASP B–17 pilots Mary Parker, Charlotte Mitchell, Blanche Osborn and Pat Bowser, now based at Las Vegas Army Air Base as engineering test pilots, were given a twin-engine C–45, gasoline and four days off to fly to Los Angeles to look for jobs at the many Southern California aircraft companies. When they arrived, the uniform-crazed tinsel city received the four statuesque women in their attractive Santiago blue with enthusiasm. The WASPs spent gay evenings at Ciro's, the Cock and Bull and the Brown Derby, where Jimmy Durante and Walter Winchell treated them like the most glittering of Hollywood stars. But at the Lockheed, North American, and Northrop factories, their reception was cold. With production cutbacks, they were told, manufacturers could not use the pilots they already had. The four Fortress pilots returned to Las Vegas, thoroughly discouraged.

Their four-engine-rated WASP colleagues, Virginia Acher and Peg Kirshner, based at Lockbourne Army Air Base in Columbus, Ohio, wrote to every airplane manufacturer and airline in the country, including a Curtiss-Wright plant in Columbus. They, too, received unanimously negative responses. A Columbus newspaper, which had followed Lockbourne's illustrious women pilots since they had first entered B–17 training a year before, was incensed. "Peg and Virginia need work," said a feature article. "They want to be test pilots. How come no job? Peg and Virginia are women. It will take a forward-looking executive to hire them. Is this a direct challenge to J. B. Davey, Curtiss-Wright general manager? You *betcha!*" But Mr. Davey was not "forward looking." None of the nation's aircraft companies or airlines were.

With characteristic enterprise, a group of WASPs at Maxwell Field, Alabama, started a newsletter in late October, as a forum for job opportunities. One issue listed twenty-four aviation companies in Alaska, and announced that the Surplus War Aircraft Division of the Reconstruction Finance Corporation's Defense Plant Corporation was planning to hire ten ex-WASPs to ferry ex-war planes from storage depots to sale points. But mostly the newsletter attempted to quell the rampant rumors about possible flying jobs in places like China and South America, although attachés at the Bolivian Embassy in Washington asserted that their country would assign any and all WASPs to flying duties—*if* each woman would bring a cargo airplane with her. By that time it was tempting to buy a war surplus C–47 and go.

Though she tried to ferret out flying opportunities for her WASPs, Jacqueline Cochran was at a loss as to what women pilots might be able to do after they were no longer flying for the military. "The future of women in aviation will not be on airlines," Cochran admitted to the New York *Times* in a November 5, 1944, article, "but perhaps on feeder lines, in aerial photography, crop dusting and instructing."

Cochran envisioned another opportunity for women pilots, "selling aviation" to the women of America. "Women are only in the market for about twenty percent of the airlines tickets," Cochran said. (In early December, she would be named a director of North East Airlines, to assist in attracting more women passengers. Thus, after all of Cochran's racing feats and the experience of twenty-seven months during which women flew the nation's largest, fastest war planes, America's top women pilots were still being relegated to airlines' marketing departments, like Amelia Earhart at Transcontinental Air Transport back in 1929.)

The WASPs, Cochran assured the *Times* reporter, appreciated their AAF training and experience, and would find some way to continue flying. "You won't keep these women out of the air," she said, recovering her spirit. "People never stay grounded once they learn to fly."

While WASPs scrambled for nonexistent flying jobs, they were by no means ignored by aviation organizations spotting a highly trained group of women. Trans World Airlines recruited them to be stewardesses, promising that as soon as women were hired as pilots, the WASPs would be the first to be called to the cockpit. WASP squadron bulletin boards bore notices for aircraft accident analysts and Link trainer instructors. The Alabama Institute of Aeronautics, proudly announcing the opening of a resident flight school for young women, asked for a WASP with a commercial license, her instrument and instructor's ratings and ground school rating if possible—to assist the student staff counselor. By far the largest recruiter was the Civil Aeronautics Administration, in the market for control tower operators and ground-to-air communicators at navigational aid stations. As December 20 came closer, many WASPs applied to the CAA, to keep salary checks coming in, and to guarantee that they at least would stay close to airplanes.

But eventually the reality hit that flying jobs were out of the question. At Las Vegas Army Air Base, WASP B–17 pilot Mary Parker drew on her physical education major at Russell Sage College and qualified for overseas duty with the Red Cross, to be stationed in the Pacific at rest and recuperation centers for B–29 crews. But her flying colleague, Charlotte Mitchell, wrote to her parents in Tulare, California, "How would you like your youngest daughter home for a spell? I suppose I can find some sort of job, but it's certainly going to be a letdown after this life." Male B–17 pilots, most of them returnees from combat, began to arrive at Las Vegas to take over the WASPs' engineering missions. The Las Vegas commanding officer allowed his four women B–17 pilots to continue flying through November. They savored every flight, testing rebuilt engines and instruments over the Grand Canyon and Lake Mead, as if it were their last. On November 30, it finally was. The four WASPs rented a cabin in the mountains and drowned their sorrows in a splurge of champagne.

In the fall of 1944, as the Air Transport Command sent its experienced men pilots around the world ferrying airplanes and flying supplies into war theaters, WASP ferry pilots based at New Castle began to get new assignments, too, interspersed with their pursuit deliveries. War planes which had seen combat duty were being shipped back to the United States to be used on AAF bases until they were ready to fall apart. Then they were flown to reclamation centers, a euphemistic term for the salvage heap. The airplanes, all red-lined, were put in marginally airworthy condition for their last flight to the junk pile, and WASP ferry pilots were given the delivery orders. If ever there was a "dishwashing" job in the AAF, this was it.

In November 1944, a group of war-weary A–24 Douglas Dauntless dive bombers had to be flown from Old Harbor Air Base, Baltimore, to the reclamation center at Reading, Pennsylvania. Nancy Batson was flight leader of the group of WASPs assigned to the mission, although she would soon learn the compass in her airplane did not work. Among her flight was pursuit pilot Jill McCormick, who expressed some anxiety about taking off in a red-lined A–24, especially since the runway pointed out over Baltimore harbor. Nancy reassured her. She had recently flown a war-weary P–38 from New Hampshire to the reclamation center at Aberdeen, Maryland. Its radio did not work. On takeoff, she found that there was no throttle tension to hold the throttles at full power while she took up the landing gear, and one of the engine cowlings was attached by a single hinge. But she had made it to Aberdeen. Jill shrugged and climbed into her A–24.

As Nancy and several other WASPs took off, Jill revved her engine. It sounded all right. Cleared for takeoff, she pushed the throttle forward and the Dauntless roared down the runway, lifting off neatly at 120 miles per hour. Jill breathed a sigh of relief as she made a climbing turn over the harbor toward Reading.

Suddenly there was an explosion. Smoke and flames burst from the engine and enveloped the cockpit. Then there was silence, just the whishing of the wind as the A–24 slowed to a halt in midair. Recalling the ditching procedures learned in pursuit training, Jill immediately pushed the nose of the dive bomber into a dive, to build up enough speed to pull out just before hitting the water. It worked. The plane skimmed the surface, spray drowning the canopy.

A crowd of air base GIs that had gathered on shore saw the canopy open and a small body rise out and start stroking away from the Dauntless, which, still smoking, hissed and upended, its tail sticking like a fin out of the harbor. By the time the rescue boat got to the airplane, Jill was already being pulled onto the shore by several of the GIs. "You could have made it a lot faster if you'd taken off your parachute," a GI said to Jill as he cut the straps of her sodden seat-pack.

"I can't even join the Goldfish Club," Jill said weakly, referring to the prestigious club for pilots who had to ditch at sea. "I don't have a dinghy tag." The Dauntless, for its overland route to Reading, had not been outfitted with an inflatable dinghy. The GIs laughed and one of them offered to accompany her to the base hospital for a checkup. Jill was grateful; halfway there she began to shake uncontrollably. Three days later, after rest and observation in the infirmary, she was back in the cockpit of another war-weary Dauntless. This time she had to make a forced landing outside of Philadelphia when an oil line broke, splattering the windshield with a thick black film. Jill landed by sticking her head out the side of the open canopy.

Back at New Castle, Jill told Nancy Batson about her adventures. "Maybe it's good we're being sent home," Nancy said, sadly. "It's not too good to be up in this kind of stuff." If her number had not yet come up, these new missions pushed her luck.

Nevertheless WASP ferry pilots across the country were busier than ever. Immediately after the October announcement that the WASPs would be disbanded, the Air Transport Command looked into hiring its women fighter pilots on an individual contract basis, as it did men civilian pilots. General Robert J. Nowland, commander of the West Coast ferrying division, wrote to General George, head of the ATC, and described the desperate situation he was in, trying to find pursuit pilots to replace the WASPs. He predicted weeks of P–51 delivery delays. All too aware that the women pilots of the Air Transport Command, especially those flying the "hotter" ships, were the prime targets of criticism in civilian aviation circles, General Arnold was prompt to order that "there will be no repeat no women pilots in any capacity in the Air Force after December twenty."

The WASP staff executive at ATC ferrying division headquarters in Cincinnati, Nancy Harkness Love, received many telephone calls from her WASP ferry pilots, often indignant and desolate cries of "They can't do this to me!" Not even her hundred experienced fighter pilots, or her twenty-five Class IV and V transport and bomber pilots with thousands of hours of air experience and prime candidates for the nation's cargo and passenger airlines, had been able to find civilian flying jobs. Yet the ATC would need their skills for months to come. Her squadron at Long Beach even took up a collection and sent telegrams to President Roosevelt, General Arnold, General George, and General Nowland offering to continue flying for nothing:

> Due to the shortage of pursuit ferry pilots at this station and the excessive amount of aircraft to be ferried the WASP Squadron Sixth F[erry] G[roup] hereby offers their services as ferry pilots on a volunteer dollar a year basis without other remuneration . . . until such time as the present necessity for pilots is alleviated and sufficient pilots are trained to replace us.

Their offer was graciously, but definitively, refused. The WASPs were heartsick. They were also angered by what seemed to be an arbitrary decision by AAF headquarters. "The inactivation of the WASPs, although not unexpected, leaves me with a bitter, unhappy feeling of having been given an important job to do, and then being rudely denied the privilege of finishing it," wrote one Long Beach WASP pursuit pilot to her commanding officers. "Not, from all outward appearances, because of any lack of efficiency, need or willingness on my part, but almost without reasonable explanation or justification."

Two and a half years before, in 1942, Nancy Love had fought all summer to get twenty-five superbly qualified women hired by the Air Transport Command as ferry pilots. With the addition of graduates from Jacqueline Cochran's training school, Love had been in charge of the welfare of over 300 WASP ferry pilots. Now the thirty-year-old veteran woman flier had no idea what she was going to do. "I have no plans," she told an Associated Press reporter on December 16. "I want to get away, and think what all has happened to me . . . and to us."

We're assembled to bid you goodbye and good luck
We all hope to see you again
For we are the class that really got stuck
The members of W–10

On Thursday morning, December 7, 1944, three years after the Japanese attack on Pearl Harbor which brought the United States into its second world war, the front page headlines of the Sweetwater, Texas, *Reporter* read, "Yanks Near Industrial Capital of Saar," "Democracy Facing Crisis in Greece, Italy and Poland" and "68 WASPs Receive Silver Wings Today in Final Graduation: 4 Generals Here." The flightline at Avenger Field was decked out with an extraordinary array of twin-engine B–25s and four-engine B–17s, which had born a host of dignitaries to Sweetwater to attend the graduation ceremonies of the Class of 44–W–10. They included General Arnold, Jacqueline Cochran, General Barton Yount, commanding general of the AAF Training Command whose WASP squadron now numbered over

700, General Robert B. Williams, commander of the Second Air Force for whom over 100 WASPs were flying tow-target gunnery training missions, and General Walter F. Kraus, chief of the Central Flying Training Command to which Avenger Field had belonged. By nine o'clock Thursday morning, the Avenger Field gymnasium was overflowing with AAF officers, families of the graduates, and WASPs from earlier classes who had flown in from air bases hundreds of miles away, to attend the final WASP graduation. At nine-thirty sharp, the Big Spring Bombardier School Band struck up the "Air Corps Song"—"Off we go into the wild blue yonder, / Climbing high into the sun . . ."—and the sixty-eight graduates of the Class of 44–10, dressed proudly in their new blue uniforms, marched into the gym. The audience rose to their feet and began to cheer.

The heads of the last WASP class swirled with memories. Six grueling months had passed since they arrived, excited and bewildered, at Avenger Field. Their first day, May 29, 1944, the Texas sky compressed itself into a dark black disk, with a rim of light around the edges, and the wind picked up so suddenly that, still in their traveling dresses, they had all been called out to the flightline to hold down primary trainers until the ground crew could haul them into the hangars. Their bare legs had turned scarlet from the pebbles whipped up by the wind. Then there was the day their physical training instructors made Flight Two run around the track in the 102-degree heat and they had lined up at the drinking fountain for great gulps of water. The next day, the entire flight had washed out in a mass agony of stomach cramps. Then there was the trainee who landed in the hospital, after inhaling too much phosphene gas during their course in chemical warfare. These memories blended with hundreds kept by the seventeen other classes of women pilot trainees over the past two years.

But 44–10's experience had been unique. After only a week at Avenger Field, they awakened to the announcemnt booming over the loudspeaker system that Allied troops had landed on the Normandy beachhead. That night, they all gathered in the gym, to pray for the invasionary forces, and listen to President Roosevelt's speech crackle over the radio. Among the most moved by the news of D-Day was 44–10 trainee, Emily Chapin. Four months earlier she had been flying for the Air Transport Auxiliary in England, and she knew that many of the American and British pilots she had met were involved in the invasion.

Three weeks after D-Day came another stunning piece of news:

General Arnold had terminated the WASP training program after the class of 44–10, because Congress had defeated a bill to bring the WASP into the Army Air Forces. They all remembered the tear-stained, angry faces of the young women who had given up jobs, sold cars and homes and paid their way to Sweetwater, to join the class of 45–1, only to learn they would never begin training at Avenger Field. And 44–10 became the "Last Class." Each month, with each graduation, they watched Avenger Field shrink from 600 to 200. Then came the letters of October 3. On November 27, after pinning the wings on the graduates of 44–9, Jacqueline Cochran had broken down in the midst of her last remaining trainees, and they had all wept together because somewhere, something was terribly unfair.

Nonetheless, the lively, determined spirit common to all classes of WASP trainees refused to flag with the last one. In fact, Avenger Field had never been in such good shape. The runways, which had been under construction during most of the WASPs' training there, were finally finished. There were two new hangars, a new gym and swimming pool, and new barracks. The oil pots were gone from night flying; Avenger Field now had electric runway lights. And the flying curriculum, which had changed constantly since the pioneer Woofteds entered Houston in November 1942, had been finely honed by Jacqueline Cochran and the Training Command after twenty-five months of experience training women pilots. It was now almost exactly like the men's in the AAF. Even the WASP's physical training was carefully geared to AAF standards. Instead of the volley ball games played by the class of 43–5 in the spring of 1943, 44–10 had received twelve minutes of calisthenics, three minutes of dual combatants, five minutes of hand-to-hand combat and thirty minutes of parachute tumbling.

But as 1944 drew to a close, into the exuberant songs of Avenger Field had crept a new bitterness. Now, along with "Zoot Suits and Parachutes" and "Yankee Doodle Gals," 44–10 also sang:

We wanted wings then we got those gol darned things
They just darned near killed us
That's for shore.
They taught us how to fly
Now they send us home to cry
'Cause they don't want us anymore.
You can save those AT–6s
To be cracked up in the ditches

For the way the Army flies
Really clears them out of the skies,
We earned our wings now they'll clip the gol darned things
How will they ever win the war?

But for two hours on December 7, 1944, they suspended any feelings of frustration to enjoy the pomp of a military ceremony held in their honor. They had earned their wings. No matter what happened in the days or years to come, they were now WASPs. They would always be WASPs. And this Thursday morning, December 7, 1944, their hearts swelled to receive words of praise from the AAF's highest commanders for their accomplishment and for the accomplishments of all WASPs.

General Yount, for whom WASPs had trained thousands of bombardiers and navigators, towed targets for fighter pilots and flight tested thousands of repaired basic and advanced trainers so that AAF cadets could fly them with confidence, lauded the WASPs' success in meeting the same high standards in training and on duty as male AAF officers.

But Yount had encountered something else in his women pilots. "It is a quality which not often is entered on a report," he told his Avenger Field audience, "one which cannot be measured merely in the number of hours flown or the number of service flights completed. It is the quality of courage—of courage in the face of danger . . . WASPs have not sought to be spared the risks" of "injury and death" Yount said. "Instead, they have filled their assignments as the occasion demanded, without thought of the glory which we accord to heroes of battle. The service pilot faces the risk of death without the emotional inspiration of combat. And the WASPs have died, without being able to see and feel the final results of their work under the quickening influence of aerial action. They have demonstrated a courage which is sustained not by the fevers of combat, but by the steady heartbeat of faith—a faith in the rightness of our cause, and a faith in the importance of their work to the men who do go into combat."

General Arnold opened his keynote address with the disarming greeting, "I am glad to be here today for a talk with you girls making aviation history." He spoke of his doubts before the WASP program, about women's ability to fly military airplanes. "Frankly, I didn't know in 1941 whether a slip of a young girl could fight the controls of a B-17," he admitted. But the WASPs had proven him wrong. "You, and more than nine hundred of your sisters, have shown that you can fly wingtip to wingtip with your brothers," he said. "The entire operation has been a success. It is on record that women can fly as well as men. . . . We will not again look upon a women's flying organization as experimental. We will know that they can handle our fastest fighters, our heaviest bombers; we will know that they are capable of ferrying, target towing, flying training, test flying and the countless other activities which you have proved you can do. This is valuable knowledge for the air age into which we are now entering."

Arnold spoke of the commendations from WASPs' commanding generals that were constantly coming across his desk. They recorded, he said, "how you buckled down to the monotonous, the routine jobs which are not much desired by our hot-shot young men headed toward combat or just back from an overseas tour. In some of your jobs I think they like you better than men.

"We are winning this war—we still have a long way to go—but we are winning it. Every WASP . . . has filled a vital and necessary place in the jigsaw pattern of victory. Some of you are discouraged sometimes, all of us are, but be assured you have filled a necessary place in the overall picture of the Air Force. . . .

"So on this last graduation day, I salute you and all WASP. We of the AAF are proud of you; we will never forget our debt to you."

The audience glowed with the sincerity of the AAF Commanding General's words. Many WASPs' eyes brimmed with tears. They were left with the feeling of a job well done, not with an opportunity cut short, a difficult and important task for the AAF Commanding General.

As Jacqueline Cochran stood up to speak, the class of 44–10 straightened in their chairs. At last, they would be awarded their wings. Cochran looked along the ordered rows of Santiago blue. "The emotions of happiness and sorrow are pretty close together," she began with characteristic informality, "and today I am experiencing them both at the same time, as well as the third emotion of pride." Cochran turned toward the four beribboned generals lined up behind her. "Seldom can one see such a group of stars clustered together—no greater honor can the WASPs receive than this." The audience applauded. "I am proud that the WASP have merited praise from General Arnold and General Yount. They think the WASPs have done a good job. That makes me happy."

Cochran thanked her WASPs for their "loyalty and good sense" in the face of all the difficulties suffered because of their nonmilitary status. And as she rationalized the end of the program, Cochran expressed the sentiments of a majority of WASPs throughout the gymnasium and others across the country. "As much as the WASPs want to help by flying, we can all be happy that our Air Forces are now so built up and the progress of the war is so favorable that our services are no longer needed." And as she addressed her loyal trainees, who had grown up admiring the brave and determined aviatrix, Jacqueline Cochran moved them deeply when she said, "My greatest accomplishment in aviation has been the small part I have played in helping make possible the results you have shown."

The audience and her commanding generals gave Cochran a standing ovation. She stood before them, her large brown eyes shining. Then one by one, she called the names of the sixty-eight graduating members of the class of 44–10. Awed by such a close brush with four generals, they approached the podium nervously. As Cochran handed each new WASP her diploma, and General Arnold her wings, the Director of Women Pilots and AAF Commanding General chatted with her. "This is Emily Chapin, General, one of the pioneers. She flew with me in England," Cochran said as she and the former ATA-girl Spitfire pilot and now WASP graduate shook hands. Emily smiled as the magnetic general spoke to her, but by the time she had joined the other graduates, she had already forgotten, in her self-consciousness and excitement, what he had said.

Finally, their wings gleaming on their chests, the class of 44–10 sang some of their frankest songs for the generals, as Arnold slapped his thigh and laughed heartily. Then the entire audience stood at attention and the Big Spring Bombardier School Band began to play the *Star Spangled Banner*. Emily Chapin tried to sing, but her voice became choked with emotion. There were very few in the Avenger Field Gymnasium that Thursday morning, December 7, 1944, who were not overwhelmed. Three years after Pearl Harbor, they were, indeed, the land of the free and the home of the brave, and they were part of the greatest allied war effort ever known.

WHEN THEIR spirits returned to earth after the spectacular and moving graduation ceremonies, the newest WASPs of 44–10 found themselves wandering around Avenger Field like lost souls. The WASP training base was reverting like a war-weary airplane to the Defense Plant Corporation. Everyone had to be off the base by midnight Saturday, but the graduates had yet to be formally discharged to go home. Rumors circulated that Jacqueline Cochran was trying to arrange for them to be given actual assignments to air bases across Texas until

December 20, so that they could serve as WASPs at least for ten days. But nothing was definite.

Early on Friday afternoon, 44–10 graduate Betty Phillips and her baymates were sitting despondently in their bay, unable to pack. Suddenly a classmate ran into the bay. "Colonel Ward wants us all on the flightline pronto—*in flying gear!*" In seconds, Betty and her baymates pulled on their flight suits, grabbed their fleece-lined leather jackets and tore out to the flightline. When they had all assembled, Colonel Roy P. Ward, a thirty-nine-year-old Texas rancher, lawyer, Air Corps pilot since 1932, and Avenger Field's final commanding officer, lined up his panting women pilots and began to talk.

"We're closing down the field, as you well know. Those PTs and BTs out there have to go down to San Angelo to be stored. I have no written orders to ferry them, but you're trained ferry pilots, you're going to have at least one flight to your credit. I'm really sticking my neck out on this. If there's one scratch on those airplanes, it'll be my career." He attempted to look sternly at the blissful faces of the young women before him.

"We wouldn't scratch jello if we were on skates," said Betty Phillips. They all ran out to their airplanes and one by one the trainers took to the skies until there was a deafening swarm heading south toward San Angelo, seventy miles away. Colonel Ward followed his WASPs in the official Avenger Field plane, a twin-engine transport called the *Fifinella,* with the Disney-designed female gremlin painted on its nose. A year and a half before, in April 1943, the class of 43–4 had flown a formation of trainers from Houston to their new home base of Avenger Field. Now the class of 44–10 were flying them away—and carefully. Not on their toughest instrument check ride had their concentration been greater than during the crucial seventy-mile ferrying delivery for Colonel Ward. An hour later they had all arrived in San Angelo. Every landing was perfect.

Colonel Ward, beaming, signed over his trainers to the operations office, and then herded Betty Phillips and several other WASPs into the *Fifinella* for the flight back to Sweetwater. As the Avenger Field commanding officer took off, he joined the members of his last class singing boisterously from the trainee repertoire that, after three and a half months as their C.O., he knew by heart. Everyone felt great relief that the airplanes had been delivered safely, without a scratch, and gratitude to their commanding officer, that he would jeopardize his career just to make them feel needed—even if they were not.

When the *Fifinella* got over Avenger Field, the WASPs yelled,

"Let's buzz the tower! We've wanted to do that for six months!" Colonel Ward laughed and dove the plane at the white wooden structure sticking up into the Texas sky. With the last defiant dip of her wings, the *Fifinella* circled Avenger Field and landed. As they all climbed out of the twin-engine transport, they gave her a farewell pat. *Fifinella* had made her last flight.

ON TUESDAY MORNING, December 19, WASPs across the country turned in their government-issue flight equipment and their WASP uniforms. That afternoon, WASPs stationed at Greenwood, Mississippi, noticed a group of officers' wives rummaging through a barrel in the middle of an open lot on the field. To their horror, they realized that the women were pulling out Santiago blue slacks, jackets and skirts and holding them against their bodies for size. After an indignant Congress had railed against the "natty" and expensive WASP uniforms, they were now being carried off over the arms of officers' wives as Army surplus rejects.

At Gardner Field, near Bakersfield, California, where for nine months WASPs had flight tested all of the basic trainers out of the repair shop for AAF cadets, the base commanding officer gave the Gardner WASPs a full parade review. As the sun was setting over the Pacific coastal range, the cadets marched by the eighteen-WASP squadron and then stood at attention as cannons boomed a salute. After the din and smoke died away, the flag was lowered solemnly and the Gardner field bugler played taps for the WASPs.

On Tuesday morning, December 19, Nancy Batson flew her last P–47 Thunderbolt from the Republic factory in Farmingdale, Long Island, over the Empire State Building and on into Newark. The next day it would be on its way to an AAF Fighter Group in England, France or Italy. Nancy would be on a train back home to Birmingham, Alabama, never to see a P–47 again.

The president of Republic had given his all-WASP P–47 squadron a farewell banquet the week before. Now Nancy and the others flew from Newark to New Castle Army Air Base, Delaware, where on their last evening as WASPs, there would be another dinner honoring the twenty women pilots of the Second Ferrying Group.

It had been twenty-six months since October 1942, when Nancy Batson had arrived fresh from Birmingham at the front gate at New Castle to join the Women's Auxiliary Ferrying Squadron. In twenty-six months she had flown twenty-eight different kinds of airplanes for a total of almost nine hundred hours in the air. In all, Nancy and her

WASP colleagues had flown over sixty million miles criss-crossing the United States and Canada, in seventy-seven different types of military airplanes—every kind of warplane manufactured in America. In twenty-four hours, however, the WASP would be history.

All Tuesday afternoon, Nancy Batson went from office to office, having papers signed and turning in her Army flying gear—her parachute, her fleece-lined flying jacket, her gun. Releasing each familiar item was like feeling a lifeline slip through her fingers.

For the evening banquet held at the New Castle air base officers' club, the WASPs dressed in their nicest dresses and suits. It felt awkward to be wearing civilian clothes again. When they arrived at the officers' club, New Castle commanding officer, Colonel Robert L. Baker, presented each WASP with a corsage. The dinner was specially prepared, with fresh fruit, chicken à la King, sweet potato croquets, chocolate eclairs, and rare French wine to enhance the WASPs' flagging appetites. Toasts were long and sometimes teary. Finally, they all decided to return to the WASP barracks and finish packing their trunks. Suddenly they heard a commotion in the lobby. A male voice shouted, "Fire! Everybody out!" The lobby was filled with smoke. The WASPs joined the crowd of male officers pouring out of the officers' club. Nancy Batson looked back at the flaming building. "Let it burn!" she yelled. "Let it burn!"

On December 20, over the NBC radio network, commentator Robert St. John pronounced a requiem for the WASPs. "Today all over the country these girl fliers will be bidding each other . . . and their male colleagues tearful good-byes. They'll be turning in their parachutes and giving some last, affectionate pats to war planes they've been flying around the country . . . and they'll have lumps in their throats as they take off their uniforms and get into civilian clothes. They'll be . . . home for Christmas. But ask any one of them, and you'll learn that they don't want to go home. I've met some of the WASPs. They are intelligent girls, sincere . . . capable. They were doing an important work . . . efficiently, well. And dangerous work it was, too. But today they go back to civilian life. I called Washington this morning. I was told that there isn't even going to be any ceremony! Just some tearful farewells . . . and one thousand WASPs go back . . . into civilian clothes, because Congress doesn't think they're any longer needed!"

In Long Beach, California, pursuit ferry pilot Jean Landis took down from the wall of her room in the WASP barracks a large map of the United States. It was covered with silver stars, close to two hundred of them, all marking places where Jean had delivered P–51s. Jean had loved the P–51 Mustang. Each time she climbed into the cockpit, she had felt as if she were putting it on, like a kid glove. Jean rolled up the map, put it under her arm, and picked up her suitcase. As she walked past the Long Beach flightline toward the air base gatehouse, sixty airplanes badly needed in war theaters sat with no ferry pilots to deliver them. But it was December 20, 1944. Jean Landis and 917 WASPs were sent home.

THEY WENT HOME to the tears of their grateful families, to the empty apartments where war-lost husbands would never return. They went home to the small towns that did not understand them any longer; to the big cities filled with career girls who had more useful training. They went to the universities and colleges where the man or woman sitting beside them was receiving a free education under the GI Bill. They went home to stand outside the fences at local air bases, just to hear the engines roar; to stand in line at the local drug store to buy a war-rationed pack of cigarettes; to stand in the middle of rooms of new houses or apartments, and wonder.

They went home to face the rest of their young lives on the ground. A few chose the bottle instead. One WASP chose death. Some went to Alaska, to Mexico, to Europe, to the mountains—anywhere to think for a while. Some went home to jobs ferrying red-lined war-surplus trainers to their new owners—but, indignant at the contrast to what they had known, stopped flying forever and looked for careers where they could find fulfillment. Some WASPs went home to accolades, but that did not matter much any more. Most of them went home to anonymity, to marriages much changed, to young men who did not understand why they turned away when the war stories filled everyone's talk. They went home to America: to Fords, Bendixes, frozen vegetables, antibiotics, and to all of the windows that looked out at the sky.

EPILOGUE

The Year of the WASP

In 1947, an ex-WASP organized a gathering of her former flying colleagues who lived in Washington, D.C. She sent out invitations, arranged for refreshments and rented a hall. When the evening came, however, no one else showed up. By 1947, two years after VJ Day, the WASPs were devoting themselves to their marriages, to their careers, to their children. Seeing one another again would have been too painful.

The Order of Fifinella, founded by WASPs based at Maxwell Field, Alabama, just before they disbanded, tried futilely to keep addresses current. Three decades went by before the women entered yet a new phase of their lives. In their fifties, recently widowed or their husbands retired, their children grown, their careers achieved, as if by an instantaneous spark, the WASPs began to seek one another again.

In 1972, the WASPs convened for a Thirtieth Anniversary reunion at Sweetwater, Texas. As they marched down the main street, the sons, daughters and grandchildren of the town they had known cheered for them. A superhighway now ran through Sweetwater, no longer a cattle crossroads, but a thriving community of 15,000. But the runways of old Avenger Field were still there, and it was just as hot as they all remembered from the days they marched, sang, studied and flew at the only all-female air base in history. The WASPs decided to meet every two years for the rest of their lives.

By 1976, the Order of Fifinella, after a concerted effort to locate lost WASPs, had an up-to-date roster of almost eight hundred names. In late October, 1976, they met in Hot Springs, Arkansas. The halls of the Velda Rose Tower Hotel that rainy Friday evening were alive with gleeful shrieks of greeting. "Murph!" "The Little Colonel!" "Demerit Meurer!" The women, now in their fifties and sixties, ran to one another and fell into tight, often tearful embraces. They would pull apart and stare into each other's faces. "Is it really you?" they would say. Some had not seen another WASP for thirty-two years. Their faces were lined now, their hair was graying and sometimes pure white. But they knew one another the instant they saw the eyes. They had not changed. They were still of uncommon brilliance and beauty—pilots' eyes.

As they packed for Hot Springs, they had dusted off old scrapbooks and photo albums. Yes, they all admitted with some chagrin, they had saved everything. As cocktails flowed, the women sat knee to knee turning the browning, frayed pages. "How young we look!" they sighed. It was hard to grasp that they were no longer these exuberant girls beaming from the cockpits of bi-planes or, parachutes in hand, mounting the wing of a P–47. Now they were university professors, photographers, judges, hospital administrators, ranchers and grandmothers. But somehow, being together again, thirty-two years seemed to disappear.

Their "flying buddy" friendships were as strong as ever. And so was the pride in what they had done as young women. Whether or not they had shared a bay or a flight during the war years, in 1976 the women clasped one another like sisters.

Laughter was the prevailing sound throughout the Velda Rose that weekend, as if the entire group shared a hilarious secret. In a way they did. History had all but ignored them, but, time and time again, they admitted to one another, that as rich and fulfilling as their lives might have been since, "Nothing, nothing has ever matched those two years as a WASP."

Friday evening the banquet room at the Velda Rose was brightly lit and filled with singing.

We are Yankee Doodle Pilots,
Yankee Doodle, do or die.
Real live nieces of our Uncle Sam
Born with a yearning to fly!

No, the WASPs had not changed. One of the women, who had learned to fly in 1928, the year after Lindbergh's solo flight across the Atlantic, had flown herself from San Diego to Hot Springs, almost fifty years later.

But something *was* different. As they sang, national news cameras rolled, and most of the Americans watching the NBC Nightly News on their television screens learned for the first time of a forgotten chapter in the nation's past. Six weeks before the reunion, on September 10, 1976, the United States Senate had voted to make the WASPs official World War II veterans. The media had taken notice. Senator Barry Goldwater of Arizona had added an amendment to include the WASPs in an obscure House-passed bill which provided Veterans Administration medical benefits to naturalized Poles and Czechs who had fought as Allied soldiers. "These women are U.S. citizens who served with our Army Air Corps during World War II," Goldwater told his startled Senate colleagues. "This group of women was shunted aside by the country they served; and it is better that we should correct a past wrong now, rather than leave the record blemished." The amendment passed.

Four days later, the House of Representatives rejected the amendment. Senate passage had sent the American Legion and other groups in the massive American veterans establishment running to their supporters on Capitol Hill. They were up in arms. No wartime civilians had ever been allowed veterans status in America. Granting benefits to the WASPs would open the floodgates to all civilian organizations —the Merchant Marine, the Civil Air Patrol, certified war correspondents, and dozens of others all petitioning the Congress, and threatening the veterans affairs budget. Nevertheless, both the House and Senate Veterans Affairs Committees promised Senator Goldwater that hearings would be held in the next Congress on whether or not to bestow the honor of veteran status on World War II's women fliers.

In Hot Springs, the WASPs launched their first full-fledged national lobbying efforts. Directing them would be W. Bruce Arnold, son of the late Commanding General H. H. "Hap" Arnold. A retired Colonel in the Air Force and national representative of the Garrett Corporation, an aerospace company, Bruce Arnold was well acquainted with the ways of Washington. But his motivation was personal. "My father died in 1950," Arnold told the WASPs. "If he were still alive, he would be doing what I want to do. He promised you militarization in 1944. I feel as if I have to finish my father's work."

Now at last, the WASPs themselves would take their own case to Congress. As the women told the television news team of their upcoming legislative fight, it seemed quixotic that a few hundred unknown women in their fifties and sixties, living in far-flung regions of the country, dared to attempt to get a bill through the United States Congress. But as the television audience looked and listened, there emerged little doubt that these women who had flight-tested damaged airplanes, flown B–17s on two engines and faced barrages of ammunition fire from untrained flak gunners were not about to shrink from opposition legislators.

As the reunion drew to a close, 1977 was exuberantly declared the "Year of the WASP." The women left Hot Springs for home, but, unlike three decades before, they were determined not to fade again from national memory. They believed they had re-emerged on a wave of a new consciousness of women's right, not only to be pilots, but to be pilots-in-command of their own lives. And for the WASPs, their time was running out.

As the "Year of the WASP" began, Senator Goldwater introduced S.247, "To provide recognition to the Women's Airforce Service Pilots for their service to their country during World War II by deeming such service to have been active duty in the Armed Forces of the United States for purposes of laws administered by the Veterans Administration." Barry Goldwater was an ideal champion. A Major General in the Air Force Reserve, he had flown wingtip to wingtip with the WASPs during World War II when he was a pilot with the Air Transport Command, based at New Castle, Delaware. Goldwater had learned to fly in 1930, and with 12,000 hours in the air in over 200 different types of airplanes, he was the Congress's most experienced pilot. Goldwater's word as to the WASPs' accomplishments and competence as pilots would silence any opposition.

To sponsor the WASP bill in the House, Bruce Arnold found Congresswoman Lindy Boggs of Louisiana, who introduced H.R. 3277. Boggs was the widow of the late House Majority Leader Hale O. Boggs, who was lost in an Alaska plane crash in 1972. The following year, she had won a special election for her husband's seat by a landslide 81 percent of her New Orleans constituents. But Boggs was no newcomer to Washington. She had served as her husband's campaign director and chief political advisor for over thirty years, and had been the first woman to chair a Democratic National Convention. Around New Orleans, she was known as "Lady Lindy."

Most importantly, Boggs typified the ideal in the House of Representatives of being able to agree to disagree. This quality would be essential. For three decades, veterans legislation had been dictated by Congressman Olin E. "Tiger" Teague of Texas. Chairman of the House Veterans Affairs Committee for eighteen years, he had ceded his post to take on another chairmanship in 1972, but he still ruled the Veterans Affairs Committee through his hand-picked successor, Ray Roberts, from a neighboring district outside of Dallas. Teague had been elected in 1946 while still in the hospital recovering from war wounds. He had been decorated eleven times, more than any other member of Congress. Throughout the House and the American veterans establishment, "Tiger" Teague was known as "Mr. Veteran." He was solidly and unalterably opposed to granting veterans status to the civilian Women's Airforce Service Pilots.

Bruce Arnold and Lindy Boggs had one ray of hope to penetrate the House Veterans Affairs Committee. The second-ranking minority member was the committee's only woman, Margaret Heckler of Massachusetts. A lawyer and an eleven-year veteran of the House, Heckler was an avid spokeswoman for women's rights. She had run successfully for her first state election, in the early fifties, when her eldest child was two. A Republican, Heckler was co-chair, with Democrat Elizabeth Holzman of New York, of the newly formed Congressional Women's Caucus.

On March 15, through Heckler's efforts, the WASP bill became the only piece of legislation in history to be co-sponsored by every woman member of Congress.

In April the WASPs opened up a national headquarters in the Army-Navy Club. With its dark-paneled lounges and its long red-carpeted hallways lined with portraits of officers, the club looked like the place where old soldiers went to fade away. But on April 15, it came alive with energetic, attractive women. WASP volunteers from the Washington area dashed in and out of a small room in the corner of the lobby. A two-foot painting of Walt Disney's Fifinella hung on the door.

The national press was captivated by the WASPs. Newspaper and magazine articles (*Viva* did a cover story in May) from around the country began to fill a large scrapbook, used in lobbying Capitol Hill. Elsewhere across the country, the WASPs were busy spreading the word and collecting signatures on petitions to present to the Veterans Affairs Committees. On a street in downtown Napa, California, WASP Pat Velut Zell encountered an ardent supporter who introduced her enthusiastically to a shopping companion as a World War II "Yellow Jacket." Meanwhile in San Francisco, WASP Dorothy Davis donned her WASP uniform from 1944 and walked the cinema lines for the popular film *Star Wars*. In two days, she collected 1,200 signatures on her petition.

An unexpected WASP supporter was the *Stars and Stripes*, a 101-year-old weekly which went to over three million active and retired members of the armed services. The paper's slogan "The Veterans Shall *NEVER* be Forgotten!" was slashed across the bottom of every issue in red. The *Stars and Stripes* hired former WASP Patricia Collins Hughes to write a weekly column on the progress of the bill, and sent complimentary copies to the entire WASP roster.

On Wednesday, May 25, the Senate Veterans Affairs Committee held hearings on S.247. It was an historic day for the WASPs of World War II. For the first time, they told their own story to the Congress. WASPs poured into Washington from all over the country. The first three rows of the hearing room audience was a sea of Santiago blue. At the time of the Senate hearings, the WASPs had almost 20 percent of the Senate and almost 25 percent of the House as co-sponsors of Goldwater's and Boggs' bills.

At the Senate hearings, however, the WASPs learned of the full force of their opposition. "In the history of our nation, the veteran has, from the time of the Revolution, occupied a special place," the American Legion testified. "It is highly prized and valuable, and is to be shared only by those who have earned it. To legislate such a grant of benefits would denigrate the term 'veteran' so that it will never again have the value that presently attaches to it." The Legionnaires felt so strongly that their representative warned Committee Chairman Alan Cranston of California that the organization could never again come to Congress to advocate veterans causes should the WASP bill be passed.

Just as vehemently opposed was the administration of President Jimmy Carter. According to the Veterans Administration, half of all Americans received some type of veterans benefits. The U.S. Code definition of who had earned veterans status—that is, those who had actively served in the armed forces—could not be breached. The WASPs, who had served only as civilians, did not qualify. The VA had sent its highest ranking woman, Dorothy Starbuck, to testify before the Senate Committee. She was not an advocate for her sex that day. To grant veterans benefits to the WASPs, she argued, would be "inequitable" and "discriminatory" to all of the other civilian groups who had served during wartime.

Soon after the hearings, Chairman Alan Cranston notified Senator Goldwater that he would not report out the WASP bill. Goldwater knew another WASP amendment was the only answer, and waited for the right veterans bill to come along.

The WASPs, however, had just gained a champion across the Potomac, in a crucial wing of the government heretofore silent on the WASPs—the Pentagon. To the post of Assistant Secretary of the Air Force for Manpower, Reserve Affairs and Installations, the Carter administration appointed Antonia Handler Chayes, a lawyer and former president of Jackson College at Tufts University. Concerned with the effects a defeat of the WASP bill would have on recruitment of women in the All-Volunteer Force and on the women who began jet pilot training with the Air Force in 1976, Chayes was determined to present the other side of the WASP issue to the President. On September 20, when the House Veterans Affairs Committee held its hearings on the WASP bill, Assistant Secretary Chayes, armed with a statement carefully researched from Air Force historical archives, was a surprise expert witness.

Once again, WASPs flew, drove and railed to Washington from all over the country and donned their uniforms, to demonstrate solidarity behind their cause. As dignified and solemn as their Senate hearings had been, the WASPs' day in the House was crackling with action. Margaret Heckler sat on the committee throughout the hearings and guided the twenty witnesses like a tough prosecutor. The hearing record presented an airtight case on the military conditions under which the WASPs served, and included documentation disinterred from trunks and drawers of WASPs across the country.

Testifying on behalf of the WASPs was William H. Tunner, head of the Air Transport Command ferrying division during World War II. "Certainly someone today can partially correct the unfairness we showed by making veterans of these women who served so faithfully and well, and with little complaint," he told the committee. His wife, WASP Ann Hamilton Tunner, sat by his side at the witness table. "I feel that those few of us still alive should be here in Washington to receive national recognition," she said, "instead of this humble plea for positive identification as a veteran."

Many of the members of the Veterans Affairs Committee had served during World War II and many were pilots. The WASPs elicited a flood of memories, and the day was filled with spontaneous applause and laughter—and lots of "hangar flying."

But after Senator Goldwater testified, Chairman Ray Roberts left the hearing room, his protocol presence no longer required. The President of the Order of Fifinella, Bernice Falk Haydu, a former WASP twin-engine utility pilot, followed him out into the hall.

"I promised you girls a hearing," Roberts said in a thick Texas drawl as he shook his finger at Haydu. "Well, you're getting it. But I promise you this, young lady. The bill will never leave my committee."

It was now October in the Year of the WASP. The women had the vociferous support of the press, were pushing a majority of both the House and Senate as co-sponsors and had an active advocate within the Carter administration. Yet the leadership of the House

Veterans Affairs Committee held their bill in an iron grasp. "Mr. Veteran" was unable to see the WASPs as anything other than a group of civilian women lending auxiliary support to the real business of the war.

On Wednesday, October 19, Senator Goldwater selected H.R. 8701, the "GI Bill Improvement Act," an important piece of legislation he knew both Veterans Affairs Committees wanted passed, to host his WASP amendment. Over Senator Cranston's objections, Goldwater triumphed.

Instead of holding an official House-Senate Conference, the usual procedure when a bill is passed in two different versions, Congressmen Teague and Roberts and Senator Cranston met for breakfast the next week. They agreed that staff members from the Education and Training Subcommittee, which had written the original H.R. 8701, would meet and work out the differences between the two bills. They also decided that the WASP amendment was out. Teague, who was chairman of the subcommittee, now had personal charge of the WASP amendment to his bill. When the compromise version came to the floor of the House in ten days, he would ask the Speaker for unanimous consent to pass the bill. With this request, he would also ask that the WASP amendment be striken on the grounds that, as the bill dealt with GI Bill education benefits and the WASP amendment granted medical benefits, it was not germain under the rules of the House. If no one objected to his unanimous consent request, and no one ever had, there would be no debate and no vote. The gavel would fall and WASPs would once again be parliamentarily castrated.

Fortunately, the ranking minority member of the subcommittee was Congresswoman Margaret Heckler. Her staff member, upon entering the first shirt-sleeve session on H.R. 8701, learned of the leadership's intentions on the WASP amendment. He immediately called Heckler, who, in turn, alerted Congresswoman Boggs, Assistant Secretary Chayes, Bruce Arnold and the WASPs at headquarters.

On Monday, October 31, telegrams began pouring into Roberts' and Teague's congressional offices from WASPs and their supporters across the country. Congresswoman Heckler, with two other committee members who also objected to Teague and Roberts' no-debate approach on the GI Bill Improvement Act, circulated a letter among committee members to Chairman Roberts calling for a full committee meeting. A majority of the twenty-eight member committee was needed. Heckler got fourteen signatures. But as such a letter amounted to a committee rebellion, no one was willing to be the crucial fifteenth. Meanwhile, Congresswoman Boggs located 125 of her cosponsors to sign a petition telling the Veterans Affairs Committee leadership it was the will of the House to vote on the WASP measure. Antonia Chayes continued her efforts to persuade the White House to change its position on the WASPs. Thanks to her, the Defense Department had openly announced its support, and the President had to decide between its opinion and that of the VA. The White House chose to remain silent. The bill was coming to the floor on Thursday, November 3.

In desperation, Bruce Arnold compiled a pamphlet which included the most indisputable documents uncovered from the WASPs' personal archives throughout the year. While doing so, he came upon the discharge of WASP Helen Porter, and by chance, read what it said: "This is to certify that Helen Porter honorably served in active Federal Service in the Army of the United States." Arnold was dumbfounded. "It says, '*in* the Army!' " he shouted, and added it to the pamphlet. On Tuesday morning, November 1, the documents were delivered to the entire Veterans Affairs Committee, including Olin "Tiger" Teague.

That afternoon, Teague announced he was willing to compromise. If the Air Force would take the responsibility of certifying that the WASPs had been *de facto* military personnel, he would support a congressional authorization for it to do so. All day Wednesday, Heckler, Boggs, Arnold and two Air Force lawyers sent by Chayes, worked on a new WASP amendment. At two o'clock Thursday morning, Chairman Roberts accepted it.

Late Thursday afternoon, November 3, as the sun was setting, Bruce Arnold and several WASPs from headquarters climbed into the House visitors gallery to watch Congress act on H.R. 8701. Suddenly they spotted Congressman Teague in the middle of the House chamber. Congresswoman Heckler approached him. "Tiger, I want to thank you for agreeing to the WASP amendment," she said.

"Margaret, I am now persuaded that their cause is fair."

Heckler's mouth dropped open. "What has swung you? What has been so influential at this late hour?"

Teague held up two pieces of paper. One was a copy of WASP Helen Porter's discharge. The other was the yellowing, fold-creased discharge of Colonel Olin E. Teague. Smiling Heckler backed away as Teague grabbed nearby Congressmen. "I'll be danged," Mr. Veteran was saying, "Will you look at that? It's the same as mine."

Teague then went to the Speaker's table and asked for unanimous consent to pass H.R. 8701, as amended, with amendments. There were no objections. The gavel fell.

Bruce Arnold and the WASPs sat for a moment, then quietly climbed the stairs out of the visitors gallery. After ten long months of pleading their case to Congress, there had been no vote, no debate, no climax, no flourish of victory. Suddenly, the "Year of the WASP" was over. "We won . . ." someone said in a small voice. They stood in the empty darkened hallways of the United States Capitol and wept.

On Friday, November 4, shortly before midnight, the Senate concurred with the House-passed compromise version of H.R. 8701. At the Defense Department, Assistant Secretary Antonia Handler Chayes would soon assume responsibility for determining the WASPs' military status and issuing them official honorable discharges to be presented to the Veterans Administration.

On November 11, Bruce Arnold assured a WASP telephoning from Illinois that she could place a flag on the grave of a World War II WASP colleague for Veterans Day. On November 23, 1977, the eve of Thanksgiving, President Jimmy Carter quietly signed veterans status for the Women's Airforce Service Pilots of World War II into the law of the land.

Index

A

G

N

O

V

W

Y

Z